I0131317

CONTENTS

INTRODUCTION

This book resulted from a lunch between two digital industry veterans who were reflecting on how far things have come in 20 years – and how much the fundamentals haven't changed.

Simon was the long-time MD and Ray the foundation content director for HotHouse, one of Australia's first digital agencies, which was formed back in the mid-1990s and was a force in the digital industry until it closed its doors in 2015.

We started out producing interactive CD-ROMs and quickly moved on to this new concept of a 'website'. We like to think we invented digital content marketing! We both worked on a Microsoft publication called Communique, that was in its time the biggest computer magazine in Australia. Soon after this we published one of the first titles on the Microsoft Network, called SexBytes.

Some of our earliest (and fondest) memories of those days include sitting in a windowless sound booth which was converted into an office, answering questions about sex that were sent in by people all over the world. Although SexBytes was based on health and science and not pornography, having the word 'sex' in the title meant that we were at the top of many search results and during the first year we achieved more page views than the Sydney Morning Herald (The anatomically correct drawings and photos of couples demonstrating different positions probably helped, as well!). The fact that we weren't professionally qualified to answer the questions didn't stop us from dishing out advice to thousands of people – sounds a bit like many social media influencers today!

HotHouse also built many of the first corporate and entertainment websites in Australia, for clients including nineMSN, Microsoft, Telstra, and McDonald's.

After a couple of years working together, we hit on the idea of combining our ideas and writing skills to publish some articles in order to build HotHouse's profile in the marketplace. Today, that would be called thought leadership, but back then we didn't need a buzzword to tell us that it was a good idea to raise the level of understanding of the Internet in the business/marketing community.

Our first articles were published in 1997, although the earliest ones we could find to include in this book were from 1998. Ray moved on from HotHouse in 1998 but continued to ghost-write the articles for Simon on a freelance basis until 2012. Our writing process consisted of getting together for a drink to canvas story ideas and angles, and then Ray would go off and research and write the articles.

Over the years, we had articles published in the *Sydney Morning Herald*, *Australian Financial Review*, *The Australian*, *B&T* magazine, *Marketing* magazine, *Ad News* and on the HotHouse blog. For a few years we also conducted regular podcasts with a variety of industry specialists (which, like HotHouse, no longer exist) which created fodder for some of the articles.

The articles covered the highs and lows of the local and global digital industry. The early tone was very positive, but the dotcom bubble was soon followed by the great tech stock crash. The industry's recovery from this was a constant theme, as fortunes were built and destroyed, websites rose and fell, and traditional companies wasted enormous sums of money on expensive but dysfunctional websites.

Later, of course came broadband, open-source content management, mobile and social. As the articles point out, what became known as social media actually existed years before MySpace and Facebook were created, including some of the first sites we built back in the 1990s, such as Manhood and Campo's Rugby World.

One of the hallmarks of the articles was the no-holds-barred approach to digital developments. We were unafraid to point out the stupidity of some players, and to make pronouncements on what we thought was important, as well as making some outlandish predictions on what was likely to happen next. That's what made us decide to re-publish the articles; when we looked through the collection, we realised that while the names and numbers had changed (when we started writing, fewer than 1 million Australians had ever used the Internet, and ecommerce revenue was in the low millions), the principles we espoused had weathered the passage of time pretty well, so we decided it would be worth re-publishing our pieces as a collection.

Of course, we didn't always get things right. What we got wrong or didn't see coming included:

- **Location-based advertising**: As we wrote back in 2000, "People will simply not put up with getting a message on their phone while they're wandering through Myer telling them about a sale in their shirt size on level four." OK, so maybe they will.
- **The resilience of traditional media:** We really thought advertising agencies and TV would be shells of their former selves by now. Although you would have to say that the effect of digital technologies à la Netflix have helped create what is now called The Golden Age of Television. And as for ad agencies, continue to watch this space.

- **The primacy of Facebook and Google:** Did anyone see that two companies that didn't exist 15-20 years ago would control the global advertising market? And that more than 1 billion people a day are checking on the lives of their friends and getting most of their news on their phone through one app? Or the international political ramifications of manipulating algorithms (can you say 'fake news'?)?
- **"Plastic will win over virtual cash" (1999):** Credit card use online is nearly universal with improvements in security, but so is PayPal.
- **"Proprietary systems and technology that dictates to the user will not survive" (2000):** Apple, we're looking at you.

On the other hand, we got a number of things right:

- **The 'evil' side of ad networks:** Like some others, we were always sceptical about relying on black-box algorithms and trusting that your ad will be viewed by who the networks claim they will be. But we had no idea of the scale of what could – and did – go wrong.
- **"Distribution, not online technology, will be the biggest stumbling block to the success of ecommerce" (1999):** And Amazon's command of this aspect of commerce is exactly why they are the behemoth they are today.
- **The customer is (and will continue to be) king:** As we wrote back in 2000, "We're moving into an era where people feel empowered and are expect to be in control. This presents a huge challenge for advertisers and marketers."
- **"Cinema will become both deeper and shallower at the same time." (2006):** Digital technology has boosted the prominence of blockbusters, but it has also increased the long tail availability of independent films.

- **"New online music services will take you beyond what you know you like and move into predicting what you will like.":** As we wrote in 2006, "If you thought Napster and iTunes represent the pinnacle of personalised music, just wait until you see what happens."
- **The continuing rise of reality-based TV formats:** Though no one could have predicted that one reality TV star's posterior would break the Internet, while another one (the star, not the posterior!) would get elected leader of the free world!
- **ROI is an important, but elusive concept:** Measuring your success is more achievable online, but the relationship between online activity and profit is very fuzzy.
- **The importance of storytelling, conversations and listening:** Content marketing is an old concept whose time has come again. Google's inexorable moves towards natural search means that telling great stories and creating helpful information will drive people toward your brand.

Despite the resonance our articles have up to 20 years later, one important thing we have learned during our time in this industry is that it is impossible to know which way things are going to go, so we have not updated our original articles – and we're not going to predict what's going to happen to digital marketing and ecommerce in the next 20 years, other than to say that the customer will continue to be at the centre of successful marketing.

We hope you enjoy peering through the window at the early days of the Internet, particularly in Australia. We've certainly enjoyed the ride.

Simon & Ray

ADVERTISING

NEW THINKING NEEDED FOR ONLINE ADVERTISING

(PUBLISHED IN 2000)

I've never been a fan of banner ads, largely because they're an old-media solution applied to a new media by people who cannot be bothered thinking about how to take advantage of the unique features of the online environment.

Web users of the world have agreed with me, with click-through rates dropping to less than 1% for all but the most compelling banners. But that hasn't stopped the proliferation of banners all across the Web, even on millions of low-traffic, personal home pages at community sites such as Geocities, Tripod and Start.com.au.

Banner ads have now been around for so long (in Internet years) that a sizable body of research on their reach and effect on brand awareness has been built up. Three of the biggest online ad agencies in the US recently analysed 32,000 online surveys and came up with five "golden rules" for online branding. Their advice was to:

- Keep it simple
- Make your logo and your banner big
- Make sure your prospect sees your banner at least five times
- Stick a person's face somewhere on the banner

These days, most of Internet advertising revenue is coming, not from full-size ad banners, but from non-standard smaller size ads. A recent AdRelevance study found that although 80% of advertisers are using full banner ads, these ads account for only 37% of ad viewership, compared to 45% for buttons.

Banner ads' domination of online advertising is quickly diminishing. Not only the piece of pie, but the pie itself could start shrinking, if a recent development aimed at putting more control in the mouse of the Web user takes hold. Metabrowsers, which have been dubbed browsers on steroids, are being trialled by companies such as DoDots and Octopus. Metabrowsers allow people to create a home page by cutting and pasting pieces of content from their favourite websites and displaying them all on a personalised home page. The advertising industry should be very worried. If people can choose what to include and what to leave out on a home page, guess what they'll leave out?

Online analyst Jakob Nielsen, commenting on the potential impact of metabrowsers on Web revenue models, paints a bleak future for online advertising. He says, "It's just a temporary phenomenon that the Web is ad-driven."

ALTERNATIVES

So if banner ads are withering and metabrowsers act as the equivalent of video zapping through ads, how can advertising work online?

One way I don't think it will work is through the use of wireless (WAP) ads. People will simply not put up with getting a message on their phone while they're wandering through Myers telling them about a sale in their shirt size on level four. People don't want to be sold, they want to buy.

Ways forward for online advertising include focusing on:

Integrated marketing: Online advertising will never become big enough to support its own industry, nor should it. The Internet is a tool that forms part of a business' marketing armoury, not an end in itself. Integrating offline and online advertising, particularly as more technologies become available to muddy the water, will become a key success factor for companies.

Sponsorship: This isn't a new concept in Web advertising, but it is much more palatable to users than pop-ups and interstitials, and done intelligently can be more cost-effective than banner advertising.

Email: Amidst all the whiz-bang technology of the Internet, the best solutions could very well be the simplest. And online, it doesn't get simpler than email. Already bigger than the postal service in most countries, email will grow from 10 billion messages this year to 35 billion by 2005 (International Data Corporation figures). As Michael Slack from research company Jupiter Communications says, "Businesses are beginning to perceive email as the silver bullet for acquisition and retention strategies."

Rich media: As broadband reaches critical mass, streaming ads and full-scale interactivity will become viable options for advertisers. Of course, critical mass is a relative term, since production costs for rich media ads will be much higher than banners, while at the same time reach a much smaller audience. Marketers will need to work out at what point producing ads for a limited audience becomes economically viable.

Rich media and other new technologies offer great promise, but their shape and their impact are hard to predict. To harness their potential, marketers will need to be as dynamic as the technologies that are driving the Internet.

MORE WORK, BUT WORTH THE EFFORT

(PUBLISHED IN 2009)

While traditional advertising in Australia followed the rest of the economy over the cliff at the end of last year, online advertising stayed strong.

It's clear that not only is online advertising becoming a more important part of the Australian economy, companies are starting to come to grips with the changing landscape. As Brendon Cropper, director of Digital Training and former director of media services company Starcom Digital, says, "A lot of people in the industry would like to be able to hit a switch and turn the Internet off, because it has created a lot more work.

"In the old days, it was just creating wants and filling them. However, people now spend the majority of their time on the Internet, whether at work or at home, and that's where they make decisions about what they're going to buy."

He says that "Distribution (of marketing messages) now costs nothing. It's more complex, but there's more opportunity. It's not just putting a message in front of them – it's more a case of developing an experience for them."

While Brendon Cropper doesn't subscribe to the view that marketers should spend a set percentage of their marketing budget online (many experts have recommended 20%), he believes companies can and should be spending more than they are at present.

RETAILING VIA SOCIAL NETWORKS

Some businesses are benefitting from the boom in online advertising more than others. New research from Europe shows that there is a strong case for including social networks in a retailers' digital marketing strategy. Twenty-three percent of social networkers post their views on specific ads in social networks, while 25% regularly forward things such as ads or links to ads to their friends, according to online researcher Metrix Lab.

Other findings from the research, which was conducted for Microsoft Digital Advertising Solutions:

- Trust leads to follow-up: 64% of consumers will visit other websites to find out more about something they've read on a friend's site
- Influence opportunities through interesting and engaging content: 60% of social networkers are prepared to put sponsored/branded content on their own Facebook/MySpace, etc. pages
- Willingness to engage and interact with brands online: 43% of consumers have visited the personal space of a brand and 16% have already had a dialogue or sent a message to a brand

According to Microsoft's RetailSpeak magazine, the research "presents advertisers with the opportunity to create new models of monetisation. With 215 million user accounts estimated to exist on social networking sites globally, advertisers can identify influential social networkers as independent brand advocates to either recommend a brand to their network or integrate it into their site.

"Based on this type of performance related recommendation, retailers can incentivise social networkers creating a new form of monetisation model." Exactly what form that model will take remains to be seen. Some companies have simply paid people to comment favourably on their product or service – if the response is not honest, it will backfire, big time.

ADVERTISING WITHOUT LEAVING YOUR SITE

There has been a lot of hype about the next evolution of the Internet – Web 2.0. Most pundits will studiously avoid trying to define Web 2.0, while others will just say it revolves around interaction. A great example of Web 2.0 in online advertising is distributed content.

What's distributed content? It's those videos that pop up while you're reading a story on the SMH or Age website. When done well, they're more interesting than what you're reading and you get drawn in.

A retro type of distributed content is the return of in-text advertising, underlined text on a page that used to simply link to an ad but now provides rich media upon rollover. People hated it when it was first introduced, but is increasingly popular in its new rich media form. Vibrant Media, one of the leading companies in the area, reports twice as many bookings as they did this time last year. Its latest reported earnings doubled to nearly US$90 million.

The next phase of what Brendon Cropper calls "rich experiences on site" is technology that enables consumers to pre-qualify themselves by interacting with an ad on a website before leaving the site to complete the transaction. Qantas is pioneering this by letting users work out flight prices and availability within an ad, before heading over to the Qantas site to complete the transaction.

Brendon says this is another step along the road to the Holy Grail of online advertising: addressable media, where all types of advertising messages can be personalised based on information companies can pull from consumers.

"It's not just putting a message in front of customers – it's more developing an experience for them," he says. "Advertising is always more effective if it's relevant."

ONLINE AD NETWORKS – EVIL, OR USEFUL?

(PUBLISHED IN 2009)

Are online advertising networks a good thing or a bad thing? It depends on who you talk to. Many experts say ad networks help large portals and small sites alike unload excess inventory, while others say the rates paid on ad networks are so low they are killing the industry.

Some people are particularly passionate in their views. David Koretz, CEO of collaboration software company Blue Tie, says that "ad networks are for idiots". He calls them "a tax on lazy publishers. They are a cancer that slowly eats away at you from the inside, doing severe damage even though you feel fine. They are a cancer that has spread to nearly every publisher, and threaten to do irreversible damage to our industry." The problem, he says, is that, "We are slaves to the short-term need to make the quarter; addicts who take the revenue boost despite the pain we will endure later."

Working under the assumption that the average premium publisher only sells 30% of its inventory direct (which is backed up by research), he accuses publishers of using 'bad math' to fill the other 70 per cent.

"If a publisher sells the entire 70% of remnant inventory through ad networks, that is equivalent to selling 0.93% more inventory direct. Less than 1%!" He recommends instead that they "Figure out how to sell an incremental 0.93% as premium by innovating, not by betting the farm."

Koretz is specifically referring to blind networks. Blind ad networks offer low pricing to direct marketers in exchange for those marketers relinquishing control over where their ads will run. Rock bottom prices (CPMs – costs per thousand page views – are measured in cents rather than dollars) are achieved through large bulk buys of typically remnant inventory combined with campaign optimization and ad targeting technology.

In a blind advertising network, advertisers get either limited or no information about what webpage their ad is shown on and where on the page it is shown. While advertisers save substantial sums of money, it's practically impossible to measure ad effectiveness.

TARGETING WORTH THE EXTRA COST

David Holmes, head of Australian marketing technology company APAC Digital, who we interviewed recently for our podcast, doesn't use the same evocative language as David Koretz, but he agrees that blind networks are of questionable value. "In my experience, they don't really work for publishers – but they like it as cream."

He said that even though the price is "ridiculously low", the CPM is not that good. Also, while you can exclude some categories of content for your ads, such as porn sites, you can't include or select which sites your ads will appear on. As he says, "Don't you want to know where those ads are going to be placed?"

He said that targeted ad networks, although more expensive than blind networks, are much better value. "Targeted networks take remnant inventory and turn it into 'excess inventory'."

This is done by adding value compared to blind networks. "You need to present a compelling difference between blind and full-price ad placement."

Targeted networks, as the name implies, give advertisers the opportunity to choose who their ad is going to target, either by selecting sites which have specific demographic data on their audience, or by using new targeting technologies. Local examples include companies like Sensis MediaSmart.

By being part of an advertising network, small websites can conglomerate and negotiate advertising deals with big companies, which is also good for the advertisers because they don't have to negotiate deals with every website they want to advertise on. They can also advertise on a large number of sites that fit their brand image and positioning.

David Holmes says that the use of targeting technology will eventually lead to the demise of blind networks, so David Koretz can rest assured that a cure for online advertising cancer is being developed.

NETWORKS EXPANDING OR CONTRACTING?

But will the cure be available in time to save the patient?

Global interactive agency Razorfish, in its 2009 Digital Outlook report, predicts that traditional ad networks will contract as competition for declining ad dollars increases. According to the report, "There are simply too many broad networks competing for the same inventory and not telling a new story."

"Agencies and advertisers will look to established publishers with reliable models to focus their investments as more scrutiny is placed on return on investment. Depending on an advertiser's goal, this might include proven performers like search, ad networks, online video and targeted media, or it could mean high quality engagement opportunities with select partners who deliver unique brand engagement. The pressure points will be on 'measurability' and 'differentiation.'"

Some pundits, such as technology trend expert Jeremy Liew, disagree with Razorfish's assessment. Liew predicts there will be more networks, not fewer, as sales execution becomes a key differentiator when blind networks become replaced by targeted networks. "Sales teams typically work best when they can focus on a set of accounts with a lot of commonality, whether demographic, industry, or geography," he says.

"This means that it will be easier (not harder) for smart small teams of sales people to start their own targeted ad networks."

LET'S GET VERTICAL

Vertical ad networks, which segment advertising based on content niches and community-based websites, are growing in use in the US, and David Holmes predicts they will become an important online advertising option in Australia before too long. "This is where advertising gets relevant again," he says.

"Vertical networks take a brand into a space of massive trust if you pick smartly." He predicts that vertical networks will change the way ads work, moving from ads as brand messages to ads as a service. This isn't a new concept – David Holmes points to billboards in ancient Rome, which had community-based messages relevant to local residents.

"That worked back then because it was the only form of advertising. Today, there's so much advertising that you don't care about it – you just switch off.

With the use of vertical networks, he says, advertising will become a drag – as in dragging people to you. People expect and welcome advertising on community-based sites, as long as it's relevant and doesn't breach users' trust.

An increasing number of experts are using the phrase 'earned media' instead of 'social media'. I think that's a great way of expressing where digital media is heading: instead of just throwing messages out there, advertisers and their agencies are going to have to earn the right to speak to their customers.

MORE THAN JUST A DISRUPTION

(PUBLISHED IN 2009)

People have been chronicling the disruptive influence of the digital media on traditional media and advertising ever since the World Wide Web first appeared. But no one has articulated the effect so comprehensively or so bluntly as Bob Garfield, ad critic and columnist for *Ad Age*.

After publishing a piece on the future of media entitled "The Chaos Scenario" four years ago, the curmudgeonly critic became a poster-boy for the doomsayers. Here was someone who'd worked closely with the advertising industry for 25 years (albeit as an observer, not a participant), someone whose livelihood depends on the success of traditional media, and he was using his column in the advertising industry's bible to declare that advertising was in its death throes.

He has now turned that column into a book – one that balances the bad news with a prescription for surviving the chaos, including examples from around the globe of companies that have successfully embraced digital marketing.

As to Bob's bluntness, in a video presentation connected to the book, he says the book has two parts, and the message of the first part is "You are doomed."

"I say doomed," Bob drolly tells the camera, "because 'totally and completely f***ed' didn't fit on the slide.'" Here is a man who tells it like it is, and he told it like it is in our recent podcast.

SUPPLY, DEMAND AND COBBLERS

"The Chaos Scenario" had its genesis in a presentation Bob made to colleagues at an *Ad Age* editorial conference. Fuelled, he says, by a night of heavy drinking at the conference, he had an epiphany while preparing his presentation.

After years of observing the rise of the digital media and the reaction to it by traditional media and advertising, he realised that "If these trends continue, it's not just a disruption, but the doom of the industry.

"It was an end of times story. It wasn't just the industry – it's also my own job. I thought, 'I'm going to lose my job – and probably earlier rather than later'....This is like the Industrial Revolution, and I'm a cobbler."

A big part of the problem, Garfield says, was the way traditional media dealt with the web when it first appeared. "In the early days, we simply moved our business model onto the web. We didn't think about the fact that the advertising dollars don't make an equal transfer."

"We ignored the law of supply and demand – if there is an endless reservoir of content (and therefore advertising inventory), it will depress the price of advertising."

Meanwhile, while media outlets increased their audience, they didn't generate any income from it. And now, Bob says, there is a growing group of consumers who "believe that 'content wants to be free'. Ridiculous! Toasters, paper towels, etc. don't want to be free – why should content want to be free?"

Regarding the plans of publishers including News Limited and Fairfax to start charging for their content, Bob says that "In some pockets, people will make paid content work. Whether they can put the toothpaste back in the tube is questionable."

He cites the American HBO model, where the pay television network has specialised in high quality content that people will pay for. That contrasts with the broadcast TV model, which has been "to throw mud at the wall and see what sticks – 90% crap and 10% quality."

Ironically, he says, the pay television model is being undermined by people streaming free video using the same cable that brings cable TV into their house. "People are now using broadband to find programs they used to pay their cable bill for, so the cable companies have been hung by their own co-axial noose."

FROM SHOUTING TO COLLABORATING

But it's not all bad news, according to Bob. Out of the chaos is coming a re-birth of marketing.

"Mass is gone, micro is the new reality. Micro offers extraordinary opportunities."

Bob says businesses need to stop thinking of customers as passive recipients of information and start treating them as "stakeholders, fellow travellers. You need to deal with people as individuals.

"Drop the megaphone – the conversation is no longer about you – and get used to aggregating a community."

"You'll be able to get intelligent loyalty and even evangelism from the crowd. You're not just forging an ongoing relationship, you're getting collaborators."

Brands, he says, offer a conundrum in the digital environment. "On the one hand, brands will be increasingly insignificant. They're a proxy for information - they signal that they will be there tomorrow, and the product will be approximately as good. In the digital world, we have an infinite amount of information at our fingtertips and we don't need a proxy, we have the info.

"But the other way of looking at this, there's so much information out there, that a brand becomes an aggregation tool – suddenly brands have a different function – they save you the trouble of sifting through everything."

Bob points out that there is an enormous amount of data on customers now available via the Internet – much of it surrendered voluntarily by people. Unfortunately, the skill sets needed by advertisers and marketers in this data-intensive environment are different from those currently running the industry.

"I don't care how creative you are, in your t-shirt and sneakers. If you're 30, you better work out what you're going to do at 35."

PREMATURE PREDICTIONS: THE CASE FOR ADVERTISING

(PUBLISHED IN 2009)

Since *Ad Age* columnist Bob Garfield has been the most vocal herald of the death of advertising (see above), much of the recent defence of advertising has been aimed at Bob and his treatise.

In a guest column in *Ad Age*, copywriter and partner of Goodby, Silverstein and Partners Jeff Goodby says he gives Bob's argument 2 stars out of 5 (admitting that, while having read the original "Chaos Scenario" article, he hadn't read the book).

He writes, "Fifty years ago, San Francisco advertising man Howard Gossage said, 'People read what they want to read. Sometimes it's advertising.'"

"The fact is, much of the internet is not paying for itself, especially in the media realm. Bob would say: Great – then die, media realm, die. I, on the other hand, believe that very soon the internet will finally wean us off the expectation that everything's free online. And it will do so through a combination of micropayment entry fees and, yes, advertising that people like."

From the Escape Pod blog, written by anonymous staff of the Chicago-based ad agency Escapology, comes the prediction that Jeff Goodby's view will be found to be right and Bob Garfield's wrong. Why? Simply because Jeff works in the advertising industry and Bob doesn't.

The blogger writes, "I for one, have had just about enough of laptop quarterbacks telling me how to do my job without ever once actually having it done it themselves."

"We in the ad biz understand that people don't like advertising. Believe it or not we get that bit. We also understand that the internet has fundamentally changed commerce and communications permanently. Everybody gets that. But to suggest the advertising industry will just go away is patently absurd.

"Advertising will no more disappear than prostitution will. Prostitution isn't the oldest profession in the world. Advertising is. The world's first hooker had to first engage in some kind of promotional activity. Purring 'Hey Ugg, you want some of this?' while slowly and salaciously raising the hem of her bearskin dress.

"Advertising was and is just a means to an end: increase sales. It will shape shift into whatever means of communication exist at any given time. It has always done so."

Simon Billing, director of strategic planning at Reason Partners, a Canadian ad agency, writes in his blog that, "The most glaring flaw in the 'death of advertising' debate is that its proponents generally conflate the effects of the internet on media and control of media channels, with advertising.... (which) is the communication of a product's commercial proposition, to its prospective purchasers, in such a way as to positively affect how they feel about that product.

"Barack Obama may have had a hell of a social media campaign, but it was his brilliant oratory that moved people. And I doubt that Churchill would have convinced too many beleaguered Brits to stay the course in the dark days after Dunkirk if he'd tweeted out his message. Technology doesn't change human nature, but it occasionally adds to the ways in which you can appeal to it.

"Will new media change the form of advertising? Absolutely. Is it bad news for advertising? Absolutely not. The real story is the exponentially expanded opportunity for creativity in advertising. Where once the media dictated the form of the message (:30/:60; full page/half page; etc.), in the future its creators will."

Simon concludes that, "The internet is the best news for advertising since Marconi invented the telly."

Meanwhile, a piece in NetRegistry says that, "Advertising has changed and evolved over the last few years. It now includes visual, audio and electronic media. In fact, if you do a Google search for advertising, you may feel overwhelmed by all the options available to you now.

"So is traditional advertising - which includes billboards, radio, television, newspaper and magazine - dead? Not by a long shot. According to one top advertising mogul, traditional advertising methods are still around because they still work. The trick is to figure out who your target market is, what they want, and how they look for that information.

"Mark Twain said, 'Many a small thing has been made large by the right kind of advertising.' If you know customers, you can spend your advertising dollars on the mediums they use to look for answers.

"Remember, if you're giving your customers what they want, they don't perceive your ads as a nuisance, they see them as a service. Traditional advertising is not dead and you can use it to your advantage if you pay attention to who your customers are, and what they want."

Finally, in an example of the "at least my traditional medium isn't declining as fast as the others" defence, BusinessWeek writer Jon Fine writes that predictions of a 6% drop in ad spending on broadcast TV in 2009 are a victory for broadcasters, because overall ad spending is expected to be down more than 10 per cent.

He writes: "You may reasonably expect, given current realities and consumers' ongoing fascination with everything digital, that this will be the year the roof caves in on network TV spending as well; at first, I did. Now I've been persuaded otherwise.

"....I'm not saying the big networks will be rolling in rose petals after this year's upfronts. But if total U.S. ad spending declines by around 10% and the networks' dollars decline around 6%, they grab market share. It's not that advertisers want to boost spending on network TV. It's that they are pulling away from other media much faster, and will continue to."

He finishes up with another classic defence: inertia.

"Who in charge of ad budgets will try radical moves today, when everyone knows a failure will cost them their jobs? This is a comfort-food environment, and no ad form is more familiar to a chief marketing officer than the good old 30-second spot. (Old maxim: No one was ever fired for buying a prime-time TV ad.) Multiply a rush to the familiar throughout a marketplace, and the result is higher prices.

"The networks maintain enormous built-in advantages. They still own their airwaves and still tightly control limited airtime. Network TV remains the cornerstone of most companies' ad plans."

IN THE WAY: THE CASE AGAINST ADVERTISING

(PUBLISHED IN 2009)

Bob Garfield may be the most high-profile advertising sceptic at the moment, but he's by no means the only one.

Guess who said this? "The advertising business is going down the drain. It's being pulled down by the people who create it, who don't know how to sell anything, who have never sold anything in their lives .... who despise selling, whose mission in life is to be clever show-offs and con clients into giving them money to display originality and genius."

Believe it or not, that quote comes from advertising icon David Ogilvy. And that was long before the World Wide Web was invented. Along with Bob Garfield, others have more recently added their gloomy predictions on the future of advertising.

In his book "Advertising is Dead – Long Live Advertising", Tom Himpe points out that 20 years ago an advertiser could reach 80% of the American population with just three television commercials, while today it takes more than 150 (I'm sure the figures are similar for Australia). "Advertising is suffering because of the sheer amount of it, the lack of innovation within traditional advertising formats, and the power that media fragmentation and technology give to consumers to tune out the noise," he says.

Eric Clemons, a professor of management at Wharton, a top American business school, wrote an article for TechCrunch earlier this year, "Why advertising is failing on the Internet", that has generated 846 comments and nearly 2,000 tweets and re-tweets.

Here are a few excerpts:

- "Pushing a message at a potential customer when it has not been requested and when the consumer is in the midst of something else on the net, will fail as a major revenue source for most internet sites."
- "The net will find monetization models and these will be different from the advertising models used by mass media, just as the models used by mass media were different from the monetization models of theater and sporting events before them."
- "The idea that content has a price and net applications should find ways to earn a profit without providing free access to other people's content gets explosive reactions; when virtual reality pioneer and tech guru Jaron Lanier suggested in a *New York Times*

Op Ed that authors deserved to be paid for their content he actually received death threats. But other models are possible."

- "It is frequently argued that the advertising industry will provide sufficient innovation to replace the loss of traditional ads on traditional mass media. Again, my basic premise rejects this, suggesting that simple commercial messages, pushed through whatever medium, in order to reach a potential customer who is in the middle of doing something else, will fail. It's not that we no longer need information to initiate or to complete a transaction; rather, we will no longer need advertising to obtain that information. We will see the information we want, when we want it, from sources that we trust more than paid advertising. "

Eric Clemons also devotes a lot of space to ways he believes companies can make money from the Internet.

Tim Leberecht, blogger for CNET and VP of Frog Design, writes: "The web, and the social web in particular, reconciles artificial scarcity with relevance, and that's why more and more branding dollars are moving online. It is the ideal forum for creating an abundance of scarce moments, thousands of small great ideas instead of one great big one. These small great ideas come to live in brief moments of attachment with customers that are personalized and truly relevant for them.

"'Advertising is failure,' says Jeff Jarvis, and he thinks 'media only get in the way of customer relationships.' And indeed, how will you make more friends at a party? Showing up with a big banner around your neck that says 'I am a great friend' or engaging in a handful of conversations with strangers, listening to their stories and detecting affinities whilst accomplishing a sense of privacy that gradually becomes intimate? Right. In the end, that's what we should be doing as marketers to build real, sustainable brand equity - creating publicity through intimacy, loyalty through decency."

William McEwen, author of "Married to the Brand" and "Inside the Mind of the Chinese Consumer", writes: "Agencies must recognize that advertising is neither the outcome nor the objective. Advertising's job is to set the stage for an actual customer experience. Then, the company's performance, the quality and consistency of its products, and its human brand ambassadors will determine the company's sustainable growth and enduring success. There is absolutely no value in making a brand promise, however memorable it might be, if a company cannot or will not keep it.

"If ad agencies want to regain their position as valued contributors to a company's success, then they must also help their clients focus on promise delivery. If agencies only care about making the promise and not about helping ensure that the promise will be kept, then they are shirking their brand-building job. It is, after all, the synergy between promise and performance that represents ultimate success -- and that's true for the agency and the client."

So is advertising dead/dying, or merely having temporary breathing problems? Time will tell.

BRANDING

SPIRAL BRANDING, OR WHY LOSING $ ON YOUR WEBSITE IS SMART BUSINESS

(PUBLISHED CIRCA 1998)

The practice of marketing is built on the premise that it's in a company's long-term interest to build the brand associated with its product or service. Assuming your product or service delivers the goods, branding, achieved by consistent repetition, is a powerful tool for affecting a customer's buying behaviour.

But if we believe in the power of branding, why do so many businesses forget all about it when they build a website and the finance department asks them to justify the expense?

In the short-term, bottom-line driven environment of the 1990s, many companies are trying to draw a direct link between the amount of money spent on developing a website and the amount of money generated by the site (usually by online transactions). That's a bit like scheduling the wedding after your first date. To make a success out of your website, you need to use it to develop a rich, lasting relationship with your customers that will give you both years of pleasure from your partnership.

ECOMMERCE VS. EBRANDING

The value of good branding can't be measured against its direct costs. Kia continues to sponsor the Australian Open tennis year after year, even though increased sales of its vehicles during and immediately following the tournament don't add up to the cost of its sponsorship. That's because Kia (and for a long time before that, Ford) recognises the brand-building value of that activity.

There's a big debate going on in the corporate sphere at the moment about the value of online branding versus making a short-term profit on an online investment. How much effort should be spent on 'ebranding' - building up an online brand - as opposed to spending money promoting ecommerce, which offers tangible returns?

The bean-counters seem to be winning the debate at present. But if you think online branding isn't as important as ringing up sales on your website, consider this: Why have the stock prices of Yahoo, Amazon, AOL, Excite, etc. soared so high in recent months? It's not their balance sheets that are attracting investors (in fact, the more they lose, the more valuable they seem to become).

They're hot because these companies have, in an amazingly short space of time, managed to build strong, recognisable brands that have stretched beyond the Internet and into the mainstream. Their brands are so strong that players from the traditional sphere such as the major media companies and booksellers like Barnes & Noble haven't been able to knock them off their online pedestal.

The benefits of ebranding aren't limited to new media brands. The Internet has the potential to be a more powerful marketing tool for a tradiational brand than sponsorship, because it gets your customers involved and interacting with your brand, at a time when their attention is focused.

As MPath CEO Paul Matteuci says, "When people visit my site, I have their undivided attention. They can't remote-conrol through commercials, they rarely head for the kitchen in the middle of their activity. In addition, I can speak to these people directly during appropriate parts of the experience. Does this sound like a good opportunity to expose them to brand messages?"

THE LUCRATIVE SPIRAL

That's not to suggest that ebranding should replace activities such as sponsorship; the fact is that it is most useful when it is combined with other brand-building activities.

Businesses which have built or extended their brand online have done it by a technique that has become known as spiral branding. Spiral branding relies on a company feeding the growth of its brand by creating a 'positive feedback loop', a term popularised by Bill Gates. He used it to describe the growing strength of Microsoft's Windows software as a universal standard for computers. The more copies of Windows that are out in the marketplace, the more companies write other software programs that work with it, which strengthens Windows' position, which encourages more compatible software, and so on.

Spiral branding relies on using a variety of media to send customers to your company to buy your product or service. The most effective spiral branding on the Web starts, ironically, offline. You promote your online service on TV, print or radio to generate interest and send customers to the Web. Then you use your website to get your customers involved and start building a relationship, which includes gathering data on them such as their email addresses. You use email to reinforce the messages they're receiving offline and remind them to return to your website.

After they've been to your website again they receive their next regular email - and on it goes.

SPIRAL SUCCESS

Media companies have a leg up on using websites to start a branding spiral because the strength of both media is in providing information. The pay TV sports channel ESPN is a textbook example. The TV channel and magazine encourage viewers to visit the ESPN Sportszone website, where they get the opportunity to get involved in things like fantasy leagues and chats with sports stars. Regular emails to fantasy players and chat fans drive them back to the website, the TV channel and the magazine. And so the spiral begins.

In Australia, the Nine Network's lifestyle programs sign off with an invitation to visit the show's website on ninemsn where they can chat with stars and guests on the show. Forum participants who register with the site start receiving regular emails encouraging them to come back to the chat and tune into next weekís TV episode.

If you don't own a TV network, including your URL on all your advertising and customer communications is the next best thing. Just make sure you give them something to do once they get there, and capture their email address. Then send them regular emails which entice them to come back to your website (However, there's a catch - you have to give them a worthwhile incentive to come back, or they'll view your email as spam and will be turned off by our brand).

The key to successful spiral branding is to make sure you use each medium to its best advantage. Television, for example, is great for creating an emotional link between customers and your brand. Your website should be used for practical purposes, such as offering detailed information on your product range and providing customers the opportunity to easily part with their money on the spot. Email, meanwhile should be used for reward and encouragement.

The moral of this story? Dare to lose money on your website. Don't expect, in the short run, to sell enough stuff on the site to pay for its development. Get in it for the long haul of building your brand.

ONLINE BRANDING A DOUBLE-EDGED SWORD

(PUBLISHED IN 2001)

Things are quietening down in the online business – and not a moment too soon. Knowledge workers have stopped threatening to leave for that stock-options-and-Friday-massages job at a dot com unless their salary is doubled. That developer who left your company to start up a new Web service and became a millionaire overnight? He's now part of the B2C (back to consulting) brigade, selling off his car lease at a huge loss and using his stock options to wallpaper his bathroom. Insane stock valuations are out and working for a living is back in.

In other words, it's a great time to move or strengthen your brand online. What do you mean, you ask? How can it be a good time to spend more energy online when so many people are crashing and burning on the Internet? It's because those who are left are less focused on the hype and excitement and more focused on making the Web work.

For all the negative talk, the fact remains that five million Australians are regularly using the Internet. While it's still not as big as the TV, radio or newspaper audience, it's growing at a rapid rate, particularly among the youth market.

Over Christmas my family was playing a game called Articulate, where a player has to describe a word so his or her teammates can quickly guess what it is. For the word "newspaper", my niece said, "something that old people read". Once I got over the fact that she considers me an old person, I realised that she's right: very few young people are reading the newspaper these days – but they're on the Web.

US research reports are showing that after years of climbing, average Internet usage time is now dropping. But that's a good sign, because it means the audience is becoming more mainstream, with more people dipping their toes in the Internet. The statistics also show that the longer people have been on the Internet, the more time they spend each time they visit.

Gerry McGovern writes in his New Thinking newsletter, "Those who are running for cover and sneering at the Internet are huddling in the past. The Internet is big. There is blood on the tracks, sure. That's a necessary shedding of youthful skin. The Internet is maturing and it's going to get bigger, badder, better, and it will blow away anyone who thinks they can live and work without it."

As far as branding goes, you have more control over your brand online. Online, the brand is the experience, not just the product. You're not relying on wholesalers, retailers and advertising outlets to put your product or service in the hands of customers. When someone buys a shirt at Target, the brand experience is tied in closely with Target. If you sell it online, you control the whole experience.

Of course, with that control comes extra work and responsibility. Display, distribution, promotion – unless you outsource all those activities, you need to develop competencies in doing all those online. It is the prospect of that extra work that is keeping many Australian brands from being very active online.

If your product or service is online-only, you have the added complication of needing to offer a uniquely compelling service simply to survive. And if mastering the intricacies of the World Wide Web wasn't daunting enough, WAP and broadband are waiting around the corner.

But that hard work is balanced by the potential of greater rewards – particularly the feeling you get climbing in the ring with your customers rather than dealing with them at a distance.

COMMUNITY: A TIMELESS TOOL

(PUBLISHED IN 2009)

Although the Internet is renowned for embracing new ideas, chewing them up, spitting them out and moving on to the next one, a few concepts keep coming back, albeit in new forms.

The most resilient of these is the concept of communities. They're in vogue again, for exactly the same reasons they have been in vogue in the past – the concept of community aptly describes the nature and the potential of the online environment.

A couple of years ago, I was asked to write a piece for the "My Five" column in the *Australian Literary Review*, where people write about five books that have influenced them in their career. The first book I listed was *Net Gain: Expanding markets through virtual communities.*

Net Gain was one of the first books that put Internet communities into a business context: not surprising, since authors John Hagel and Arthur Armstrong were both McKinsey & Company consultants. They argued that the web is used for three basic purposes – to find content, to interact and to complete transactions – and they correctly predicted online communities would grow around these purposes.

The book was strongly criticised by purists when it was published; they said that on the Internet there was commerce and there was community and never the twain shall meet. Not long after this, of course, eBay appeared. eBay is the textbook example of the type of virtual communities predicted by Hagel and Armstrong: a vibrant, self-regulated community where people barter and sell, and make friends (and dollars) along the way.

They also forecast the disintermediation of industries such as travel, describing a future that looks a lot like travel sites such as Zuji, Wotif and TripAdvisor.

Hagel and Armstrong focused on the economics of virtual communities, arriving at the unsurprising conclusion that companies that begin developing online communities before their competitors have a significant loyalty advantage.

Well, I look around and it's like *Net Gain* all over again. Then, we had Geocities and Tripod. Now, we have Facebook, MySpace and Twitter. And companies with thriving brand communities are weathering the economic storm better than those without.

ONLINE COMMUNITIES = GLOBAL VILLAGE

Online communities are the present-day embodiment of the concept of the global village, first touted by sociologist Marshall McLuhan in the 1960s. Many pundits predicted that television was the technology that would bring to life McLuhan's vision of people connecting with each other regardless of where they live. But TV was never going to make that connection, because it's all one-way traffic (they don't call it 'broadcasting' for nothing!).

The Internet, on the other hand, gives people the opportunity to seek out others who share an interest, philosophy, job or anything at all – without any regard to geography. They can dip into information created by traditional media on a topic, but they can then register their own reaction, or go off and conduct their own discussion with others totally separate from the original piece that started the conversation.

The challenge for marketers is to harness the power of communities. In our recent podcast, info architecture, social media and online community expert Stephen Collins from AcidLabs talked about the rise of brand communities.

He stresses the importance of companies looking at the big picture when thinking about online communities, saying, "You need to look at all things that draw people in (to your brand) on and offline, everything should be aimed at cultivating community on an ongoing basis."

Stephen points to the example of Harley-Davidson, which rebuilt the entire company around owners of Harleys, who have always had a strong offline community. In fact, the company's community-focused business strategy started back in the mid-80s, long before the World Wide Web existed.

Plenty has been written, in publications such as *Harvard Business Review*, about the work Harley Davidson has put into its community strategy. It founded the Harley Owners Group (HOG) with the aim of growing customer loyalty, enhancing the 'Harley-Davidson lifestyle experience', and bringing the company closer to its customers. With a free one-year membership included with the purchase of every new Harley, HOG now boasts more than one million members who gather offline and online to ride, make friends, and participate in community service

Stephen Collins says Harley takes a holistic approach to its brand community. "Riders who get together each weekend are also meeting online.... It's an incredibly powerful approach that influences everything Harley does."

A key element to successfully leveraging brand communities is to recognise that a community strategy needs to be integrated with the overall business strategy; it's not just a marketing strategy

"A lot of what Harley has done is altruistic," Stephen says. "What they're doing is first and foremost good for the community, it's not just about their business. This builds immense passion around the brand."

COMMUNITY COMES FIRST

It's important to remember that the brand is built around community, not the other way around. "Smart brands reward community members, and not necessarily with money – it might be giving them special designations or positions within your brand community. These things mean a lot to these people."

He warns about entering into financial arrangements with community members. "If you're hiring community members to do things for you online or offline, it has to be carefully managed, so those people aren't seen as puppets for the brand."

Microsoft is a company that used to do a great job of blending online and offline communities for its brand, creating advocates that weren't on the Microsoft payroll. "As a customer, you would spend 12 months communicating with someone like Nick Hodge, Robert Scoble or Frank Arrigo via an online forum, then you go to a conference and meet them, find out they're a good guy – it takes it all up a notch."

"It's not a technology thing – less than 10% of this is about the technology," Stephen says. "What's much more important is the grass roots effort to turn communities that already exist informally into cohesive brand communities."

Stephen Collins has four pieces of advice for organisations that want to get started on brand communities:

1. Be open to new ideas.
2. Be prepared to fail frequently.
3. Find the right people.
4. Give them the skills to build the community effectively.

Finally, companies embarking on this effort need to realise that brand communities "are a long-term thing – you've got to do it for years to make it successful."

THE LINK BETWEEN PERSONAL AND CORPORATE BRANDING

(PUBLISHED IN 2009)

As the name implies, social media is about people – not just your customers, but also you and your colleagues. As such, it's as much a personal branding tool as a corporate branding one.

When you set up a Twitter account to build relationships and (gently) talk about your company/product/service/event, you're also building your own profile. If you're writing for your corporate blog, you and your company are in a symbiotic relationship that benefits both of you; if you broadcast your company connections on your Facebook account or are the admin for a Facebook fan page, you're blending your corporate and personal image on several levels (which is why a lot of people choose not to do it).

The most obvious social media tool where this blending occurs is LinkedIn. LinkedIn has variously been dubbed an online resume, an online contact list and Facebook for business people. The idea of forming networks to advance business aims online is a powerful one, and hundreds of millions of people in hundreds of industries in more than 200 countries have bought into the concept.

Unlike most businesses, LinkedIn has experienced enormous growth during the global financial crisis as redundant or worried individuals join to extend their reach in the search for possible employment opportunities.

It's also being used by more and more businesses as they encourage their senior employees to sign up (following the lead of junior employees who've already registered) and connect with potential customers (as well as potential future employers – such is the double-edged sword of social media).

31

LINKING TIPS

The top social media experts use LinkedIn as a key component of building their personal/corporate brand online.

Chris Brogan, who has built a vicarious interest into a career as a strategist and commentator on social media and whose blog is ranked in the top 10 of the Advertising Age Power 150, says that the most important thing to remember when putting together your LinkedIn profile is to think about who's going to be reading your profile.

"The first horror show I see when reading other people's LinkedIn profiles is that they're written completely dry, as if robots are the only thing that will read them. Though one should write with (search) robots in mind, this is still a human network, so write as if you want someone to actually read your profile."

He goes on to say that, "Make sure that when people read your job description, they are thinking about how to put you to work on their issues. I state my company's primary functions in the first sentence of my current role, so that people can see what I'm bringing to the table alongside my own personal skills. Thus, my job description states what I'm doing, but also what I can do."

CONNECTING – STAY LOCAL, OR GO BIG?

LinkedIn's official position on building your network is that you should only connect with people you know well personally. Chris Brogan says, "You're welcome to take their opinion on that. I've chosen to accept with anyone who connects with me, and I've only had to drop one person ever for abusing that connection.

"In my view, expanding my network means that you will find the person you need by searching through my network, and that I, at least in theory, can help you get to the person you need for your business efforts.

"Your mileage may vary. I will do it my way, as most folks who connect with me eventually come calling to reach someone else that I've added, and I feel good every time I can be helpful."

Guy Kawasaki, a former Apple executive who has also built a successful career as a social media pundit (132,000 followers on Twitter (editor's note: 10 years later this was more than 1.5 million) and a top-rating blog), agrees, writing that, "By adding connections, you increase the likelihood that people will see your profile first when they're searching for someone to hire or do business with. In addition to appearing at the top of search results, people would much rather work with people who their friends know and trust."

HELPING NETWORKS

So how do you conduct conversations online? It's not really different from doing it face-to-face.

"If you know how to network, you know how to social network," Lori Gama, owner of DaGama Web Studio recently told a US newspaper.

She said the online realm is all about creating relationships and the "help thy neighbor" mentality. Gama believes up to 90% of the time spent interacting on Twitter, Facebook and LinkedIn should be "dedicated to helping and listening to others", leaving only 10 percent for shameless self-promotion. She said others "will appreciate the help that was offered in the past and even help with the promotion side."

Someone else who has taken the idea of using LinkedIn to help others and elevated it to an art form is Chuck Hester, communications director at email marketing services company iContact, based in Raleigh, North Carolina.

Chuck, interviewed for a recent podcast, has, like Lori Gama, taken the 'social' in social media and attached it to the word 'good'. A natural networker in the face-to-face world, he uses LinkedIn to help people without immediate thought of gain and encourages people to 'pay it forward' and help others rather than reciprocating to the person who helped them.

As Chuck explains, "I live the pay it forward lifestyle - if you are helped, don't help that person, instead pay it forward to someone else in need. The blessing will come back to you!"

On LinkedIn, the blessings have come back to Chuck in the form of 9,000 connections, he's now published a book about the philosophy: Linking in to Pay it Forward: Changing the value proposition in social media.

According to Chuck, it's important to make sure you use tools like LinkedIn to meet people face-to-face wherever possible, or, as he puts it, "put the social back into social media".

He's used his network to set up dinners in every city he visits (including a recent visit to Sydney) and has set up the LinkedIn Live Raleigh networking event, drawing on his 1,500 LinkedIn connections in a city of less than a million people (it's like having 1,500 connections in Adelaide) to run networking events regularly attended by more than 350 professionals.

His advice for businesses coming to grips with social media tools such as LinkedIn, Facebook and Twitter is to "remember that it's not about the tool, it's about the person behind the tool – social media is about talking one to one."

His top tips:

- Spend time understanding what the different tools are, what they look like. Just hang out in these communities and watch what goes on for a couple of weeks before jumping in.
- Be transparent." You can't sell yourself, you need to be very subtle about that aspect."
- Be helpful. "You need to come into social media saying 'How am I going to help you?', as opposed to 'How can you help me?'".
- Don't push any agendas. This may sound counterproductive to conducting business through social media, but as Chuck points out, "The social media highway is littered with the bodies of companies who have tried to sell through social media and have failed miserably in the process."

It's not rocket science, but it's based on good old-fashioned politeness and embracing the true meaning of the word 'community'. In the long run, it will improve your business through relationship-building, but just as importantly, it's also a great way to help make the world a better place.

Especially in these tough times, that's got to help you sleep better at night.

BRANDING'S MID-LIFE CRISIS

(PUBLISHED IN 2009)

I've been on my soapbox for a long time about the need for advertising and marketing to focus on delivering commercial value and not rely on vague concepts like "brand dialogue".

While I've found a number of pundits who agree with me, I didn't expect to find one who came from the heart of branding territory. But Jonathan Salem Baskin, former executive at Grey Advertising in New York and Los Angeles, has become, in his words, "a defrocked priest, a turncoat" and is now working as a marketing and strategy consultant, as he puts it, "propagating my heretical views" for the same types of clients for whom he previously produced branding campaigns.

When I read Jonathan's book *Branding Only Works on Cattle*, I knew I had to do a podcast with him. The book is a quip-filled expose of branding, and he talks like he writes.

His mantra (which was the original title of his book) is "Branding is dead; long live branding." He's not against the concept of branding, but he's an opponent of where advertising has taken branding during the past 50 years.

Today, when most people think of a brand or branding, they think of value attached to a logo, or a feeling about a company, its product

35

or service. But Jonathan says that's all wrong, arguing that "Brand is behaviour. It's an aggregation of what people do in regard to your product. What they *think* and *feel* is important, but what they *do* is paramount."

"A brand is not a feeling," he says. "Feeling hungry is not the same as being fed."

Jonathan traces the decline of branding back to the 1960s. While a generation of advertising pioneers such as David Ogilvy created ads that sold benefits, he writes, "The post-war boomer generation started to come of age, and individuals and society alike became more critical, self-conscious and self-focused.

"A distinction between branding and marketing was codified, and the distance between the two functions grew. Branding took charge of imagery, creative and humour, all in the hope of overcoming consumers' growing mistrust, while marketing continued communicating product or service benefits."

Now we have a situation where the advertising industry trumpets the values of brands via annual top 10 lists, but can't say how that "value" translates into sales and profits. Jonathan says measuring the value of brands is like trying to prove that ghosts exist. "Brands are ghosts that everyone senses, but nobody sees."

Meanwhile, thanks to the Internet, markets and marketing have changed. "We are the last generation of the mass media market," he says. As predicted in *The Cluetrain Manifesto*, markets are now conversations, happening both with and without companies involved, and recommendations are trumping traditional advertising.

But, as he points out, this isn't a revolutionary new development so much as a return to the pre-branding era. "If you take out the technology contexts, it's like the medieval markets – a return to the way we used to buy. It's a messy village of communication, a return to the old-fashioned – we need to unlearn the habits of the past 50-60 years."

According to Jonathan, "People don't want advertising; they want help to make the right decision." In the same way, he says, "People don't want to visit doctors, per se; they just want to feel better."

He's not necessarily a big fan of current online marketing trends, pooh-poohing most of the attempts by corporations to use Facebook, Twitter and blogs to build a relationship with their customers. "I don't want a relationship with you – I want you to buy the damn thing, over and over."

He encourages his clients to build behavioural models around broad context – "a pathway to buying stuff" – and not around a relationship with the brand.

Just as customers are seeking more practical solutions to their needs and wants, Jonathan Salem Baskin has some very simple, frank advice for marketers and advertisers in the modern marketplace: "Make a declarative statement – 'Our shit is better than their shit' – and then deliver on that promise.

"Replace brand image with brand reality. Identify what consumers want – good service, better financing, new features – and go about delivering it to them."

THE MUSIC IN ME: HOW BRANDS MERGE WITH ENTERTAINMENT TO ENGAGE CUSTOMERS IN THE DIGITAL ERA

(PUBLISHED IN 2010)

Music is a powerful tool. It can affect people emotionally and physically. Hearing a few bars from a familiar tune can immediately take you back to a time long past, when you first heard that song or when something significant happened while the song was playing.

It's no surprise that sharing music has emerged as one of the most popular ways people have used the personal realm of social media.

As most people know, music was one of the first media to be transformed by the Internet – think Napster and Kazaa. But just as illegal file sharing led to iTunes, music has moved from a grassroots phenomenon online to one that is becoming entwined with big brands.

And just as the music industry has been transformed by a bunch of computer rebels into the most sophisticated content business in the world, the relationship between musicians and big brands is transforming the way these big companies do their marketing. They're operating on the musicians' and fans' terms, quietly building relationships rather than beating people over the head with their brand message.

In a recent podcast, I spoke with Andrew Reid, strategy and research director of Peer Group Media, a company that blends art and commerce through the connection of brands with music and entertainment.

Peer Group owns a street magazine publisher (The Music Network) and runs a creative agency, a sponsorship agency, a PR consultancy, music industry research and talent management company. It holds the rights to a number of high-profile music festivals and even hosts a cutting-edge art gallery on the first floor of its building in funky inner-city Glebe. "We have one foot in industry and one foot in the branding/marketing space," according to Andrew.

Peer Group represents the growing trend of brands using music to identify with their customers.

"Engagement is the new 'e-word', like e-commerce and e-business," Andrew says. "The question more and more brands are asking themselves these days is 'How do we engage people with their love of music?'"

Well, it's not enough to just sling your name on a festival as a sponsor. The music community thrives on social media, and as with all social media, a brand needs to be authentic and talk with, not at, customers, and listen to what they have to say.

Brands need to find "tribes" that closely match the demographic they want to attract, and then they need to become part of those tribes by offering tangible support to their musicians. By consistently doing this, they will eventually gain the support of those tribes.

Support means things like identifying new bands and helping them to get off the ground. Andrew Reid says it is a long haul. "A multi-year campaign is needed – you have to be prepared to get involved in the community.

"You might want to just test the waters, but if you test and get it wrong, there are implications. The brands that succeed in this space are those who are committed to the long-term."

Adopting music as a branding strategy is not without its risks. "It's a high wire act – the rewards are great, but so are the risks," Andrew says. "You need to be sensitivity of the creativity of artists. This is truly a merger of art and commerce, not a takeover. It's a reality check – you are dealing with creative people and sparks will inevitably fly."

Andrew says the most surprising insight he's found since getting involved in music marketing is the connection between the 'dark side' of online music – file-sharing – and the commercial side. While not condoning the illegal practice, he said his research shows that file sharing can lead to commerce. "Illegal file sharers are actually an integral part of the industry. They begin by sampling, but they move on to buying.

"Exposure to an artist is doing a lot for that artist. You need to balance how much money is being lost vs. how much exposure an artist is getting."

Mind you, bands like the Grateful Dead worked that out years ago. They made no effort to stop people taping their concerts and producing bootleg recordings, secure in the knowledge that more people were getting exposed to their music (which wasn't the type of music that was getting much airplay on major commercial radio stations), and that when those listeners wanted a higher quality recording they would buy their albums.

If you've ever met a "Dead head", you know how much of an influence musical artists can have on someone's life. For brands who want to put the effort into it, there is a powerful vein of association to be tapped.

THE CASE FOR EVIDENCE-BASED MARKETING

(PUBLISHED IN 2010)

One of the big buzz phrases in healthcare these days is "evidence-based medicine". In a nutshell, it means relying on a combination of the best research evidence along with the experience doctors have treating patients as a way of achieving the best outcomes when diagnosing and managing illness.

Serious medical research began only in the late 19th century, when healthcare professionals started to collect scientific evidence on the effectiveness of bloodletting, a technique that had been used for centuries to cure a multitude of ills. Of course, once the evidence was collected it was determined that transfusions, the opposite of bloodletting, were a much more effective way to save lives.

Well, a South Australian marketing academic has compared modern marketing managers to medieval doctors, and he's doing what he can to take marketing out of the dark ages.

Professor Byron Sharp, Head of Ehrenberg-Bass Institute at the University of South Australia, has spent his career collecting evidence to support (or refute) traditional marketing theories, and his results stand many traditional beliefs on their head. His new book, *How Brands Grow*, lays out these theories.

I was put onto Byron's book recently by a friend and colleague who said to me, "You've got to read this book – it makes sense for once!" I found it refreshing and interesting, and after reading it I decided to interview Byron for our a recent podcast.

IT'S NOT 'DIFFERENTIATE OR DIE'

In his book, Byron takes aim at many traditional marketing assumptions. He says most marketers have based their beliefs on untested theories, in the same way doctors believed in bloodletting. "Doctors worked using their impressions, assumptions, commonsense, accepted wisdom and scattered bits of data," according to Byron. "This is similar to the working practice of marketing managers today."

"The marketing equivalent of humoural imbalance theory (a discredited theory of the makeup and workings of the human body adopted by Greek and Roman physicians and philosophers) may be the Kotlerian (editor's note: Philip Kotler, who coined the 4Ps marketing framework, is known as the Father of Modern Marketing) 'differentiate or die' world view, where marketing success is entirely about creating superior products, selling these at a premium price, targeting the most likely buyers and advertising to bring people's minds around to the product's superiority."

Byron's work at the Ehrenberg-Bass Institute has been to, as he puts it, "support research into repeatable patterns." He compares research into marketing today to the study of chemistry in Victorian era, though instead of Florence Nightingale and Marie Curie, the data is coming from Nielsen-type data. It's setting up, as he says, "Scientific laws for the social sciences, to cut down on the mysticism around business."

One of the fundamental principles that a careful study of marketing research has uncovered concerns the role of brands. Brands, Byron says, are used to simplify our lives, providing heuristic (experience-based) shortcuts to make decisions on our purchases.

We know some brands, he says and keep buying them, as long as we are reminded by advertising. Advertising's main role is not to persuade, according to Byron, but to "refresh memories, so we're not thinking too much, and to signal to people things like 'this is expensive'.

"Ads are not changing your mind, but catching your attention, to nudge and remind memories that are starting to fade."

TARGET AT YOUR PERIL

Byron Sharp also takes aim at the current emphasis on target marketing, particularly as the use of the Internet by consumers makes it harder to get broad reach and easier to reach niche audiences. While targeting, is useful, he says, "Thou shalt target goes too far. More targeting is not necessarily better."

He says research has shown that the 80/20 rule, where the top 20% of your customers account for 80% of your business, is fallacious. He says the true figure is more like 60/20 – the top 20% of customers account for just over half of your business. Under those circumstances, he says, "Not talking to 80% of your customers is ludicrous."

He also warns against relying on a specific person as a brand's target market. "Talking about your customer as one person is not sophisticated – it's the dumbest way to market. Adopting the idea that your customer is one person, of a particular gender and age, is dangerous thinking."

Byron holds up the example of Burger King in the US (branded as Hungry Jack's in Australia), which focused nearly all of its marketing effort recently on its most important target group – 22-year-old men, who were already visiting Burger King seven times a month.

They succeeded in lifting share with that group, but their edgy advertising, such as the Subservient Chicken website, anti-diet rappers and a creepy costumed Burger King character, alienated all its other customers. Meanwhile, during the same period main rival McDonald's continued to target all age groups, and lifted their overall market share at the expense of Burger King in other demographics.

(Editor's note: The podcast associated with this article) contained significantly more of Byron's thoughts on marketing – in particular, how online search advertising fits into this research-based framework.)

CONTENT/ CONTENT MARKETING

NET USERS ARE MEDIA JUNKIES

(PUBLISHED IN 2001)

If people are spending more and more time on the Internet, what else in their life is missing out? Do they treat the Internet simply as another information and entertainment medium and adjust their other media activity to squeeze their Internet time in, or do they spend less time doing other activities?

There is a big debate going on among media pundits in Australia, the US and Europe as to whether - and how much - Internet usage draws time away from traditional offline media such as television, radio and print. One theory goes that Internet users are so busy surfing they don't have time to watch Friends or A Current Affair or read the daily paper.

For instance, the US Audit Bureau of Circulation recently reported a broad decline in newspaper readership for most of the nation's largest newspapers. At about the same time, another US survey revealed that 40% of all Net users read newspapers online.

In Australia, meanwhile, www.consult reports that 57% of all Internet users are watching less TV, and 30% are spending less time on the phone, while radio use has felt the least impact.

But it's not nearly as clearcut as it seems. The eStats research group reviewed and evaluated relevant research on the Internet and media use and came to the conclusion that Net usage is not in any significant way displacing traditional media usage.

It uncovered studies that show Net users actually read more magazines and newspapers than their non-Net user counterparts. They also listen to more radio and, according to a recent BJK&E media study, watch 6% more prime time and 21% more late night television than non-Net users. This is corroborated by the Arbitron NewMedia research study which found that the heaviest Internet users reported somewhat higher TV viewing levels than lighter Internet users. Further, increases in Internet usage were associated with increases in the perceived importance of television.

eStats' theory is that Internet users, who tend to have a higher income and more education than the average citizen, are media junkies. The Internet is simply another means of getting the news and information they want, and it gives them more control. It, complements, rather than detracts from, traditional media use. A Find/SVP study adds that 73% of Net users regard the Internet as very or somewhat indispensable.

According to the marketing newsletter ICONOCAST, "Our media plate is so full now that some users are resorting to drinking latte to speed up their reading, but the two most frequently mentioned solutions are increased scanning and less TV viewing."

ICONOCAST surveyed its readers about their media use habits. Although this was not a scientific sample, ICONOCAST explained, "While most subscriber studies present a capsule view of a larger market, we feel that due to an acute need to monitor about 35 Internet marketing-related sites, newsletters, lists and publications, Net marketers represent 'the bleeding edge of the leading edge.' Put another way, this survey is a preview of things to come."

When readers were asked how they cope with information overload, many indicated that their media consumption habits were in flux. One adaptation is an increase in article scanning. Many readers have also become more selective by filtering or disregarding content.

Besides article scanning (mentioned by 16.7%), ICONOCAST subscribers watch less TV (16.8%), and read fewer consumer magazines (13.3%); fewer computer magazines (12.6%); fewer newspapers (11%) and fewer books (10%). Some even sleep less (9%), while 8% report that their social life suffers. Several macho readers wondered "what information overload?"

If you want to look at a larger group that offers a preview of things to come, examine the statistics on young adults. According to the Australian Bureau of Statistics, nearly half of all 18-24-year-olds accessed the Internet during the 12 months to May 1998. And the anecdotal evidence is that much of this growth in use is coming at the expense of TV viewing.

According to Australian media researcher David Keig, media fragmentation is at its most marked among 18-24 year olds. "They have grown up with computers around them and computer games can be one of their major leisure activities," Keig says. "Many have embraced TV for movies, sport and music. And all of this is displacing their free to air TV viewing. Even though they are yet to be hostage to family demands, their time is not limitless and media has to fight for its place in their lives alongside work or study and social activities."

The Internet is just a fact of life for this group, not something to be feared. The appeal of the Internet, according to Keig's research, is that there is an anarchy about the Net that ties in with their lives. "They don't want to be eternally spoon-fed by traditional media. They're carving out their own lives and tastes and have a keen nose for what's cool and what isn't cool.

"On another level, the Internet merely extends and embraces what they've always done - playing with the computer, chatting, following interests, digging out information, etc. The Internet just makes this all so much easier, more immediate and more accessible.

"On yet another level, the scope and flexibility of the Internet allows them to create their own world through it that mirrors their lives, interests and activities. Sometimes, with chat rooms, it's like going to a pub or a club. At other times, it's a parallel to magazines or a real and attractive alternative to yet another episode of Blue Heelers.

"This is not to say that their free to air TV viewing is doomed. Favourite programs will keep their place in their lives. But much else that was previously watched just because it's on will be replaced by the new, be it pay TV or the Net."

So where do the 18-24s go when they're on the Net? Interaction - meaning email and chat - is an important feature. In Australia, ninemsn is banking on the future of chat, offering more than 300 hours of chats per week on subjects ranging from computers to sex and relationships. Despite all the ready-made content from ACP and Channel Nine available to ninemsn, chat appears to be the backbone of the network's push for Australian Internet traffic.

According to international Internet research company Relevant Knowledge, sites rating high with the 18-34 group (it doesn't break them down further) include communities such as AOL, Tripod and Geocities, Pathfinder (the Time-Life network), Ticketmaster and Hotmail. Young men spend lots of time at technology sites ZDNet and c/net, as well as adult sites, while young women visit Whowhere (a person search site), Amazon, weather sites, city sites such as Sidewalk, comedy sites and general interest women's sites.

Since half of this age group, a lucrative present and future consumer group, are already online, you can't ignore the need to use the Internet to reach the youth market. As David Keig says, "To reach this age group one should not give any one medium primacy over any other. They will switch from free to air TV to pay TV to the Internet as naturally as older people switch from watching TV to reading a magazine. A total view of their whole media and technology consumption, in which the Internet is on the same level as traditional media, is needed to reach them effectively. "For 18-25-year-olds, convergence isn't just some abstract notion, it's already happened. Their free to air viewing may be declining, but their total 'screen time' is not. It's just that there are now different types of screen to spend time with."

CONTROL COMES – AT A PRICE

(PUBLISHED IN 2001)

One of my favourite Dilbert cartoons has Dilbert sitting in the office of his pointy-haired boss. On the wall behind the boss is a chart with a large oval in the middle and lines spraying out connecting to smaller ovals. "We're decentralising our operations to achieve flexibility," he tells Dilbert.

The next panel is labelled "18 months later". The two men are sitting in exactly the same positions as before; the only difference is that the chart behind the pointy-haired boss shows a traditional hierarchical pyramid. "We're centralising operations to achieve economies of scale," he tells Dilbert. A thought bubble above Dilbert says, "The man is a management genius."

Nearly all businesses, regardless of their size, go through cycles of centralising and decentralising. When it comes to websites, the current global trend is toward centralising resources to do as much of the work in-house as is practical.

Where three years ago companies would go to a web developer, ask them to build a website, and then go back to the developer periodically to update it (or, more likely, leave it until it was obsolete), today they hire teams to maintain and re-develop their site. Many companies have moved onto the next stage: They're looking for tools that will allow them to automate the process of updating and creating new web pages.

This trend has given rise to a new and lucrative software business – content management.

A content management system is an application designed to help organise and automate collection, management and publishing processes online. You need a content management system when:

- you've got too much information to publish by the old coding method with the team you have in place
- your information needs to be updated more quickly than your current team can deliver
- you want to use bits of content on more than one page or site
- you need to separate the design of your website(s) from the content so that as design changes each page of the site you don't need to modify each page by hand.

Since the field of content management is so new, there are no universally accepted standards for what content management systems are or do. They started as in-house applications designed to automate the process of updating a site. They have evolved into multi-million dollar commercial applications that allow companies to operate a whole series of websites with a minimum of technical resources.

BUSINESS CONTROL

Why should marketers care about content management? Here are three good reasons:

- A good content management is essential if you want to make full use of personalisation techniques. Rules-based personalisation, where you serve up particular bits of content such as articles or ads depending on a user's demographics or their journey through your site, needs a content management system to feed those bits of content.
- In the same way as it works with personalisation, content management allows you to get full use from traffic measurement systems, allowing you to track how visitors travel through your website.
- Most importantly, an efficiently-functioning content management system takes the day-to-day control of your website away from the computer jockeys and gives it to you. You're not waiting days for the over-worked web developer to change those three words on one page; you can get information about your new campaign up in hours, rather than weeks.

Give business users control over their website, no longer at the mercy of a few geeks and the CIO. Sounds great, eh? Well, unfortunately it's nowhere near as easy as it seems.

As David Walker wrote recently in the *Sydney Morning Herald*, "Just five years ago, it was almost impossible to waste a million dollars building a website. But modern, 21st century Internet technology means that any medium-sized organization with Web ambitions can now pour a seven-digit sum straight down the hole almost instantly. And one of the easiest and most efficient ways to do this is to buy the wrong Web content management system."

When you decide to go shopping for a content management system, he writes, "you'll find yourself like a shopper in some exotic bazaar, besieged by a hundred CMS vendors all promising to fix your problem. And whichever solution you choose, you're a good chance to end up out of pocket and unhappy."

APPLES AND ORANGES

Why is it so easy to go wrong with content management? A big reason is that because content management is such a new field, there are plenty of traps for young players (and let's face it, at this stage we're all young players).

One of the biggest traps is getting locked into a highly proprietary system that requires add-ons and customisations so you're forced to buy expensive licences and hire specially-trained systems integrators every time you want to add or change something on your system. To avoid spending millions of dollars for the privilege of updating your own site, you need to select a content management solution that integrates with other software and systems.

It's difficult to rate the various content management solutions because few of them really compete head-to-head. Some focus strictly on content management, while others include personalisation, search, measurement and anything up to a complete enterprise resource management (ERM) solution.

This leads to another issue. If you purchase a system that strictly does content management, you need to make sure it is compatible with your company's IT infrastructure. In other words, you can't make a decision without getting your IT department involved (better still if you can get them to pay for it). But if you get too deeply mired in the technical issues, you defeat the purpose of getting a CMS, which was to avoid getting bogged down in the technical side of things.

Tackling business process issues is equally important. A CMS forces you to examine the workflows you go through in creating, approving and publishing material on the Web. This is a good thing, but developing those workflows can be a painful process, particularly if your current methods are undocumented and ad hoc.

And did I mention how expensive content management is? David Walker's comment about throwing a seven-figure sum down the hole is no exaggeration when you include extra hardware requirements and internal and external implementation costs, particularly as the major content management vendors are US-based companies and the Australian dollar continues to head south.

And what do you get for your cool million? Not anything that guarantees you a smooth run. The top rating independent analyst Forrester Research gave to the top content management systems on the market was only three out of five.

If content management systems are expensive, hard to compare with each other and painful to implement, why has this turned into a multi-billion dollar business? Because, as the size of the average corporate website climbs past 10,000 and towards 100,000 pages, the ability to automate processes is better than the alternative – trying to manage it all by hand. This is one process that is growing too large to be outsourced, no matter how much of a genius the boss is.

A BET EACH WAY ON EMAIL MARKETING

(PUBLISHED IN 2002)

A lot of debate has taken place on the HTML vs. text-only format question, and almost all commentators agree on one thing: if you're

51

asking whether HTML or text-only is the one to use, you're asking the wrong question.

David Hallerman writes in eMarketer, "The question remains why a marketer would want to employ one e-mail format over another—or even if the marketer needs to make such a choice."

Jeanne Jennings, writing in Clickz, tells marketers, "Text or HTML? The short answer is both. You should offer your email newsletter in both formats. Let recipients choose which they want to receive."

Hallerman concludes, "Marketers should always offer readers the choice between text-only and HTML e-mails. And even the HTML e-mails need to be carefully crafted (from a tech side), since not all e-mail software reads HTML e-mails alike. Making such distinctions means more work, but the traps avoided can mean the difference between e-mail fiasco and e-mail success."

LOOK BEYOND THE FORMAT

Once you've made the decision to cover your bases and produce your email newsletters in multiple formats, you can concentrate on the really important bit – getting the content right. As Mike Adams, the self-proclaimed "Email Doctor" (he claims to have invented the first personalised email marketing technology for the PC), says, "It doesn't take long to realize that effective email marketing is really about the human side of communications, not the technology side.

"Technology just gets your message to them and allows you to handle the administration of it all, but technology doesn't do anything to connect your customers' interests with the products, services or ideas you're offering. It's like television: TV technology is meaningless, it's the message that comes across the airwaves that matters. "That's why, in your email marketing efforts, you should move beyond the technology issues as quickly as possible and focus on the human interaction. Let the medium of email fade away, and spend your brain

CPU cycles thinking about how to connect with the human beings on the other side of the screen."

Kathleen Goodwin, CEO of iMakeNews, agrees. "Think like a publisher," she says. "Content is the magnet that will drive your readership and attract subscribers. Great content equals great readership."

Goodwin recommends that marketers should learn to "look beyond traditional measurements of click-through and open rates and, instead, observe how people interact with and read content, promotions, and fact-based information.

"By tracking customers' interactivity with content and using new customer segmentation techniques, savvy marketers now produce smart newsletters customized with messages based on true reader preferences and complex customer intelligence."

Information in email newsletters should speak to the interests of audience segments, according to Goodwin. "Stand in your customer's shoes so you can understand her individual interests and behaviors. Then, individualize and personalize communications and incentive offerings accordingly.

"Your content should continually evolve as you get to know each customer better. Over time, you'll become so adept at meeting readers' needs, they will seek out each newsletter – and click links that build their brand knowledge. Even if you need to send 10,000 different newsletters with each 'mail drop,' this approach yields deep rewards."

LET'S GO TO THE VIDEO

(PUBLISHED IN 2005)

The fastest-growing trend in new media is to make it look and work more like TV. Now there's a new idea!

Online advertising is booming in terms of both dollar growth and market share around the world this year.

So what's driving the growth in usage and the corresponding growth in advertising dollars? One of the biggest reasons is the increase in broadband take-up, which is allowing website owners – and advertisers – to increase their use of video.

The more the Internet resembles mainstream media, the more comfortable the bulk of the population feels with new media. For traditional agencies accustomed to TV, online video makes the Internet more familiar.

Video ad growth is at a tipping point. An eMarketer spokesperson told *Ad Age*, "Television and the Internet will find ways to complement each other; winner-take-all is not the name of the game."

VIDEO ALLURE

Website publishers are only now beginning to beef up their video content offerings enough to accommodate ads. As of earlier this year, only about 30% of all US online publishing sites supported streaming content, according to Advertising.com research quoted in the eMarketer report.

The lack of standardization among video and video players on Web sites is also an issue. "Anything that forces people to download a player before using it makes it take that much longer," the eMarketer spokesperson said.

BusinessWeek reports that a key difference between TV advertising and Internet video ads is that, "In many cases, Web surfers knowingly click on ads. The ads promise entertainment. It is this model – tracking consumers and enticing them – that gives Internet video much of its allure.

"Not all advertisers are pulling back on TV. But most of the big ones – in technology, cars, and consumer goods – are finding ways to reach viewers online. Using everything from five-second bursts to short films, they're shaking up the nature of video ads, which have been stuck in a 30-second rut for decades."

Online video ads can be inexpensive to produce – often only a fraction of the cost of making a standard 30-second TV spot. In Australia, Carlton Draught's "Big Ad" (if you're one of the handful of people in Australia who haven't seen it, check it out on YouTube), which was produced originally for the Internet and was later shown on TV, is the exception to the rule, with a rumoured six-figure production budget. It has now been viewed by more than two million people online since its release earlier this year.

Popular video ads such as the "Big Ad" are emailed by the thousands (it started by Fosters sending it to their 4000 employees and encouraging them to pass it on to friends and family), giving advertisers a free boost, whereas on TV they pay huge fees for each airing.

The current trend is to use video ads online as extensions of TV campaigns. The way it works is that TV broadcasts the message, while the Internet is used to follow it up, tracking down customers in their online niches, measuring their behavior, and – if possible – luring them into online showrooms.

BusinessWeek reports that American Express has claimed its online traffic grew by 31% last year when it ran a Jerry Seinfeld ad on its website and then promoted it on TV.

It quotes John Skipper, an executive vice president for sports network ESPN: "Five or 10 years from now I wouldn't want to be just selling TV commercials."

CHANGING THE RULES OF THE GAME

However, despite the fact that the types of ads and the size of online advertising budgets are increasing, there's still just one line item for online advertising, without regard for the various available interactive opportunities, Greg Smith from Carat Fusion told delegates at the recent Ad:Tech conference in New York, which was covered by the ClickZ website.

"We paint in colour, but they evaluate in black and white. A lot of what we can do doesn't fit into the current evaluation model," Smith told the conference.

Because agencies can't put online advertising into terms that traditional advertisers are familiar with, it's hard for interactive to get a fair shake at measuring up, Smith said.

Palisades Interactive VP Arthur Chan, who also spoke at the conference, said, "We're running into the problem where interactive is seen as not being as efficient as spot TV or spot cable. Any time you give the client an opportunity to buy TV, they'll do it."

Chan said, "We have to change the rules of the game. There's work being done now in the industry to get rid of reach and frequency and replace it with reach and engagement. You have to go deeper than CPM."

Smith recommended that all agencies ask themselves a couple of questions to determine whether they are using interactive marketing to its fullest. For creatives, the question is "What would you do creatively if you couldn't do advertising?" For media people, the question is "What would you do online if you couldn't advertise online?"

"Once you give people permission to think that way, once you start talking in that kind of language, we're at a whole new level," Smith said.

TO BLOG, OR NOT TO BLOG?

(PUBLISHED IN 2007)

Blogging is big business. Blog search company Technorati is tracking 70 million blogs across the world (more than 250,000 in Australia), and the company estimates there are 175,000 new blogs being launched daily (Editor's note: at time of publication, Tumblr alone has nearly 400 million blogs.)

A growing number of those blogs are being used by businesses, to communicate with their own employees and, increasingly, to connect with their customers more deeply than was ever thought possible.

At its most basic level a blog (contracted form of 'web log') is an online journal or diary that is updated by its owner. The blog owner writes about whatever possesses them, with the most recent entry appearing at the top of the web page and other entries following in reverse chronological order.

Obviously, blogs are well-suited for individuals who want a platform for their views, or who want to keep their personal network informed about their lives. But more and more businesses are using blogs to present a more personal face to their company, to gather feedback from customers, fix small problems before they become major ones, and explain the company's view on various issues.

Corporate blogs can be written by anyone from a CEO (the CEO of General Motors gained positive notoriety last year for writing quite frankly about issues facing the company as it slid towards bankruptcy) to marketing executives to technical engineers (even the notoriously-controlling Microsoft has encouraged its employees to create their own blogs and comment on others' blogs).

Debbie Weil, US marketer and author of *The Corporate Blogging Book*, says, "Any attempt by execs at the top of big organisations to think out loud and listen to feedback from employees and others is laudable."

When I say that more and more companies are using blogs to communicate with their customers, I should qualify that to say that more and more companies *internationally*. For whatever reason, corporate blogging has not taken off in Australia.

As technology writer Graeme Philipson wrote recently in the *Sydney Morning Herald*: "You'll look high and low for a web log penned by an Australian captain of industry.... I can't find a single decent Australian corporate blog. There may be some within corporate intranets, but none presented to the outside world."

It's not because there isn't a demand for it. Philipson quotes business blogger Wolf Schumacher, who recently asked his readers if Australian chief executives should be blogging. Nearly half the respondents agreed with the statement "Yes, I want to know what they stand for", and another quarter with "Yes, isn't it the task of executives to communicate with people?" Philipson wrote, "It's not a terribly scientific poll but it does make you wonder."

For those of you who are considering getting into the untapped market of Australian corporate blogs, here is a primer on why you should do it, what to do and not to do, and who's doing it well (I'd like to acknowledge the contribution of blog experts such as the aforementioned Weil and Schumaker as well as Backbone Media and B.L. Ochman (say that quickly), author of *What Could Your Company Do With a Blog?*).

WHY BLOG?

- Blogs make putting information online supremely simple.
- Blogs help your business set new agendas.
- Blogs are quicker and cheaper than traditional PR and marketing.
- Blogs can help your company achieve greater visibility through search engine marketing.
- Corporate blogs are an effective way to communicate with and get feedback from customers, prospective customers, the public, employees and business partners.
- Blogs allow you to respond quickly to negative trends or rumours.
- Blogs are an excellent way to stay in touch with customers and hear concerns that can be an early warning system of potential problems.

BLOGGING DOS

- A blog has to be unedited and free from the arse-covering disclaimers that will rob it of credibility.
- The people who represent your business in your blog need to be the same people you would allow to represent the company to the media or to the outside world.
- Your blogger should be someone who is a good/excellent writer, with a sense of humour.
- Write about something you are passionate about – it doesn't always have to be directly about the business.
- Surprise, surprise - the more often you update, the more popular your blog will be. Shorter, more frequent updates are better than sporadic long entries.
- Make sure you link to other blogs. Linking to other blogs shows your customers that you are involved in the community.
- Make sure you have an email link or contact page. Sometimes readers might want to contact you, but not leave a comment on your blog. If you're worried about getting your email address spammed, a simple contact page can prevent that.
- Make sure you give customers the opportunity to leave comments. If a blog is used only to share information but does not allow communication between the company and its customers, its success and popularity will suffer.
- An RSS feed is a good way for people to read a blog without always actually visiting the blog. You are more likely to pick up regular readers if you have an RSS feed.
- Mine the 'blogosphere' for insights on what's important to your customers – as well as monitoring what is said about your business.

- Build relationships with bloggers (but not the wrong kind – see blogging don't's below).
- Identify your key influencers and help them achieve their wants and desires.

BLOGGING DON'TS

- In the words of B.L. Ochman, "Nobody will read a blog that sounds like a press release or standard company crap. You have to be willing to be conversational, sometimes controversial or confrontational, and occasionally intrepid."
- Don't attempt to buy off bloggers. In the US, Edelman Public Relations, working on behalf of its client Wal-Mart, approached the writer of the Consumerist blog, who had been criticizing the company, and offered to pay him to stop bagging the company. Not surprisingly, the blogger exposed Edelman and Wal-Mart's offer on his blog and the resulting publicity backfire was huge.

A COUPLE OF EXAMPLES

- *Fast Company*: This magazine, once the bible of the new economy but struggling for relevancy since the tech crash, is publishing insights and comments from its writers that have been edited out of their articles, giving readers more than what they get in the (sanitized) print publication.
- Bernie's Blog: Character blog for an organic food company aimed at younger customers and written by "Bernie, the Rabbit of Approval". Simple, yet effective.
- Jonathon's Blog: Graeme Philipson writes that this blog by Sun Microsystems CEO Jonathon Schwarzman "is the archetype of

what a chief executive's blog should be. The ponytailed president writes clearly and succinctly, and presents his views on various matters that affect his company and the industry. The blog might be seen as a soft marketing tool, but it has a loyal following (about 4000 hits a day), and its success helps promote Sun's image as a caring, sharing corporate citizen."

Although there are many business advantages to corporate blogging, it's not for everyone. You need to have a culture of customer focus for it to work. But then, don't we all need that?

French Internet researcher Cedric Firmin writes that one of the key characteristics of corporate blogging is that it "considerably modifies the way companies and partners interact. Thanks to corporate blogs, we should – if everyone plays the game – witness real 'win-win' situations: better products for the customers when companies take into account their comments; bigger market shares for companies owing to the customers' loyalty. To put it simply, corporate blogs are good for good companies and bad for bad companies."

WHEN EVERYONE'S A PUBLISHER... WHO'S GOING TO READ IT ALL?

(PUBLISHED IN 2009)

Back in the 1980s, when the development of personal computers and desktop publishing software made it easy to change page layouts without expensive typesetting, it was said that anyone could become a publisher. Yes, it did increase the flexibility and reduce the cost of putting out a publication, but you still had to sell enough copies to pay the printer, the truck driver and the distributor – there was still a sizable cost risk for independent publishers.

But today, with the advent of the Internet and free blogging tools, the phrase "everyone's a publisher" has now come to fruition. Millions of individuals and small independent publishers create their own news, comment and videos to share with friends, family and the rest of the world.

Not surprisingly, this has had a devastating effect on traditional media outlets. With people splitting their time between TV, radio, magazines, newspapers and now millions of websites (not to mention the reduced spare time all these self-publishers have to consume media), TV audiences are splintered, radio is slumping, and print circulations are steadily dropping. Newspapers have been particularly affected as one of their major sources of revenue – classified advertising – migrates to the web at a cracking pace.

AUSTRALIA - HANGING ONTO EYEBALLS

But, as with the global financial crisis, the situation for the traditional media is not as bad in Australia as it is in the rest of the developed world. Consumption of traditional media and associated revenue, while dropping across the board, hasn't fallen off a cliff the way it has in the US and Europe, with newspaper and magazine closures tossing more than 25,000 journalists out of work in the past two years.

And while media proprietors in those countries decry the rise of the Internet, Australian media companies are making the transition to integrating their online and offline presences with a fair degree of success.

There were nearly 2 billion page impressions per month to the top three media web portals (Fairfax, News Limited and ninemsn) in the year to June 2009, up more than 30% from 12 months ago. And while I have my own questions about the measures used to compile these numbers, and the revenue from those impressive statistics still doesn't compare to the rivers of gold from their traditional activities, at least the Australian media giants are maintaining some semblance of reader loyalty.

So what's different about the Australian market? I asked Andrew Jaspan, former editor of *The Age*, when interviewing him for a recent podcast.

Andrew puts it down to geography. Our continental isolation historically kept publications from outside our borders gaining a foothold in Australia, while the distance between our major cities limited our choices for local news.

Even before the Internet took hold, Andrew points out, Australian newspaper companies enjoyed a duopoly in Sydney and Melbourne and a monopoly in Brisbane, Perth and Adelaide. Contrast that to the UK, where readers can choose from up to 15 newspapers on the newsstands. Fewer information choices translates into increased loyalty when choices increase.

THE FUTURE IS SMALLER

But will even the relatively-well-placed Australian media companies be able to survive the massive shift in revenue that's underway, with smaller players and Google joining the fight for what is likely to end up being a smaller pie?

The answer isn't clear. But one thing that is clear is that they will need to change the way they operate. Andrew Jaspan says this is already happening as savvy news outlets remove the barriers between online and offline reporting staff and let subject specialties, not media specialities, determine who covers a story.

The next step is much harder: admitting that media consumption habits have changed and reducing the spend on dead trees and printing presses.

For content providers who have a foot in both the online and offline camps, their offline offering will shrink in conjunction with shrinking audiences and revenues. Look for daily publications to contract to weekend-only, focusing on analysis, commentary and investigative journalism rather than breaking news. And without the economies of scale associated with daily publication, they will have to become more expensive – e.g., newspapers will cost up to $4 an issue rather·than $1-$2.

This isn't a new concept. It was first proposed by Barney Kilgore, the visionary publisher who turned the *Wall Street Journal* from a stock report sheet read by 30,000 financiers in the 1940s to a financial media behemoth with more than a million readers by the 1960s.

The Internet wasn't even invented when he issued his prediction; he was worried about the immediacy impact of radio and TV on newspapers. But with the added impact of millions of 'always on' channels, his view of the future of media is now coming to pass.

THE BUSINESS OPPORTUNITY IN ONLINE

Media commentator BoSacks recently left a comment on an article in *Time* that claimed that "Content is rapidly being devalued". He wrote, "The author keeps referencing the decreased value of content.... perhaps it is his company's content that is devalued, but not content itself. That may be splitting hairs, but somebody is always going to make money on content. Right now it is Google; next year or in ten years it will be somebody else. Google doesn't make the content, but they sure as heck have figured out how to make money on the content. So it still has as much value as before, but perhaps not by the same companies. Get over it and start thinking in 21st century terms and actions."

The unlimited bandwidth of the Internet is demanding to be filled, and as well as creating opportunities for independent publishers, it's also creating opportunities for marketers. The phrase 'everyone's a publisher' applies as much to businesses as it does individuals.

There is an opportunity – I'd go so far as to call it an obligation – on companies to deliver content to their customers.

People are desperate for trusted content to help them make practical decisions in their lives. While they're not looking for advertising in the traditional sense of the word – which creates problems for companies, ad agencies and traditional media – they're turning to news and information sites, friends (through recommendations and comments), online advice (ditto) and to companies themselves.

Joe Pulizzi from the Content Marketing Institute says that "Almost all marketing tactics are optional today. The one strategy that is not optional is developing ongoing content for your customers to help them make better buying decisions. Organizations and individuals around the world are trying to figure out how to create valuable and relevant content. While there is plenty of bad content out there, great content is in high demand.

"....(Content) is needed and accessed by consumers now more than ever. Pure media content is still wanted and needed, but there is more competition today, and since the traditional media business model is threatened, quality corporate content is becoming more important than ever (corporations have the money to invest in content)."

Think of it as a 21st century spin on the soap opera concept. Instead of soap companies sponsoring a program, they're commissioning their own online content to support their brand.

This creates yet more competition for traditional media and their advertising-sponsored business model – unless they can transform their business model in a way that can capture this market, through a new type of partnership with businesses.

Most marketers probably find the prospect of becoming content creators quite daunting. But search engines and search terms are driving people to companies for information, and you need to be where your customers are.

YOU'RE A PUBLISHER NOW

(PUBLISHED IN 2010)

There's been a lot of talk about how the evolution of the Internet is bringing about the death of traditional media and advertising. There's also been a lot of strong disagreement about this prediction (admittedly largely from people who have a vested interest in the status quo).

I think there's a lot of merit in both views, and the truth is somewhere in the middle. The disruption to traditional forms of communication is undeniable, and the implications are getting stronger every day.

At the same time, death is such an extreme, loaded term. Just as people predicted the death of books when newspapers (and then movies, and then television, and then the Internet) came along, and then predicted the death of radio and movies when TV came along, followed by the death of TV when videos appeared, history (and logic) tells us that decline, but not death, is on the cards.

So if we're going to be relying less on print ads, TV and radio spots, even online banner ads and buttons, how will companies get their message across to consumers? According to Rebecca Lieb, US vice president of the digital research and publishing outfit Econsultancy, we are in the middle of a fundamental shift from advertising to marketing.

Rebecca, who was a guest on our recent podcast, says that not only has online search technology made it simple for customers to connect with businesses, the evidence shows that most searchers are going straight to a company's website for more information about their products. In other words, it's not advertising driving people to your business online, it's search.

This is where the decline of advertising and traditional media comes in. Instead of relying on finding out about products and services via ads sandwiched around and between newspaper and magazine articles, TV and radio shows (or even, for that matter, on websites), consumers are increasingly sidestepping that and getting straight to the source by searching on Google.

The official term for this is a seven-syllable mouthful – disintermediation. It literally means not using mediators to complete desired tasks. And traditional media and advertising agencies have built their entire business model on serving as intermediaries, which is why they're scrambling to stay relevant in this new environment.

EVERYONE'S A MEDIA COMPANY

So what are the implications for businesses in the 21st century? Basically, it means that the jobs of intermediaries such as the traditional media are being thrust upon companies themselves. In other words, as Rebecca Lieb says, "Brands are not just businesses; they're now media companies."

There has been a fair bit of discussion on this topic, but that debate centres on shifting the media from large media conglomerates to small niche publishing businesses that are single-sponsored or 'owned' by corporations with an interest in the niche subject area.

But that still assumes that people will be consuming media from specialised media outlets. I think what we'll find is that search engines will cut out the middle man and take people straight to companies, without a media outlet (owned or otherwise) in the middle.

FRESH CONTENT NEEDED

Meanwhile, the rise of universal search – the inclusion of news, video and commerce results along with standard text content results – means that as well as making it more important to be on the first page of search results, it's also more difficult to get and stay there.

So how do you do it? Rebecca, who as well as her publishing and consulting work has written the business best-seller *The Truth About Search Engine Optimization*, says that search engines "like frequently updated content." She also points out that gaining links to your pages from other websites boosts your search results and "fresher, brighter content is more likely to be linked to."

Companies need to shift their thinking from an emphasis on advertising to an emphasis on marketing and content creation. She says that means there's "lots more media to play with. And it's free – but that doesn't mean you can mess with it."

So what do you need to do as a result? Rebecca Lieb says marketers need to change their mindset from short-term to long-term.

Marketers, she says, are used to thinking in terms of campaigns, which have a beginning, a middle and an end, then on to the next thing. She says most marketers struggle with ongoing marketing. "For example, look at most company newsletters: they have a great first issue, and then they've shot their wad."

CONTENT STRATEGY NEEDED

In the digital age, she says, you need a long-term perpetual strategy. To be able to successfully develop and execute a perpetual strategy, according to Rebecca, "You need to think like an editor."

Your web presence is now, to use one of my favourite phrases, "the beast that must be fed". To channel customers to your business, you need to have a content strategy.

That means things like introducing an annual editorial calendar for creating new content for your site, publishing regular features and what Rebecca terms "sticky stuff, like a horoscope or a crossword puzzle in the newspaper."

Many companies that are embracing this shift are hiring ex-journalists that have made the transition from traditional media to this corporate marketing-driven information revolution.

It sounds more expensive, in terms of personnel and content creation, but the money to fund this activity is matched by a reduction in advertising spend as that option becomes less useful.

Rebecca says that in the long run, however, companies will spend less on marketing overall. "Just look at the money Pepsi saved this year by not running their multi-million dollar Super Bowl ads. They spent an enormous amount on social media, but less than they did on straight advertising in previous years."

But the traditional media based their business model on a (mostly) clear separation between advertising and content. What happens when the 'advertiser' is also the content provider?

In the digital context, according to Rebecca Lieb, "Being authoritative is more important than being objective – though transparency and disclosure are incredibly important.

"If, for example, you're a sporting goods company and you publish information on your site about mountain climbing. That information can be entertaining. The information is not invalid, as long as you know where it's coming from."

Rebecca concludes: "The rules aren't different; it's the channels that are shifting."

DIGITAL MARKETING

THE APPRENTICE HIGHLIGHTS CRISIS IN MARKETING

(PUBLISHED IN 2005)

(Editor's note: This article does not really deal with digital, but we've included it for its political topicality!)

I fail to understand the current popularity of reality TV. The personality dynamics that occur when you throw 16 people in the jungle for a month and get them to vote each other off the show might have been interesting enough to watch once, but for 10 series? And at least the scenery changes in *Survivor*, whereas any shot ever taken of *Big Brother* seems to consist of a bunch of primary school drop-outs sitting around on a lounge talking about absolutely nothing.

If I see another show about minor Australian "celebrities" getting fit or fixing someone's house, I'm going to scream. Next thing you know, they'll be joining the circus. (What's that – they *are* joining the circus? EEEEEEAAAAAAUUUUGGGGHHHH!)

There is one exception to this rule: I have been riveted by *The Apprentice*. I thought I would hate it – a bunch of self-important MBAs trying to out-sycophant themselves in front of the abominable Donald Trump, the king of self-important wankers. Despite that, I decided to watch the first episode – and I have been hooked ever since.

A bunch of (mainly young) MBAs, lawyers, entrepreneurs, marketing executives, project managers and real estate salesmen are divided into two groups and try to outdo themselves in a series of business tasks against a glamourous New York backdrop. In many ways it is the antithesis of *Survivor* – the teams stay together in a luxurious penthouse at Trump Tower rather than in bedraggled lean-tos on separate desert islands – while the personality clashes are the same no matter where you put a group of Type A-personality Americans together.

I think what I like most about *The Apprentice* is that each episode is a textbook lesson in what not to do in business. In fact, in a case of life imitating art, many universities such as Harvard are now running courses for their business students where they analyse episodes of *The Apprentice* in class.

The would-be apprentices delegate too much or too little, they quibble and pout and focus on personalities rather than the task at hand, and worst of all, they consistently fail to think big. Time after time their solution to maximizing sales of lemonade, coffee, ice cream or Big Macs is to stand on the street handing out flyers and calling out to passersby encouraging them to try their wares.

When the losing team gets dragged into the boardroom to confess their sins to "The Donald", the spotlight is turned on the project manager for that task. And when the project manager is asked what went wrong, he or she invariably blames the person who was responsible for marketing. The show has been running for three seasons now and I struggle to think of an episode where they got the marketing right.

This is a sad indictment of our industry.

CUSTOMERS DOING THE WORK FOR MARKETERS

(PUBLISHED IN 2006)

OK, hands up everyone who has received a funny joke, story, link or video in their in-box and thought, "Look, I normally don't do this, but this is really funny!" and then forwarded it to some of their friends or family.

Now, keep your hands up if you do that *every day*. You may be surprised how many people still have their hands up.

According to a report recently released by US interactive marketing firm Sharpe Partners, 89% of adult Internet users have done it, and a staggering 25% say they do it every day. In total, 63% of adults say they do it at least once a week, and three-quarters of people answering the survey share their emails with up to six other people.

Not surprisingly, humorous emails are the most common type shared with others (88%), followed by news items (56%), health/medical items (32%), religious/spiritual (30%), games (25%), sports/hobbies and business/finance items (24%), and "sexually provocative" content (12%). (I'm sure if people were totally honest on the survey that last one would have appeared a bit higher on the list).

The good news for marketers is that the study found that adding overt brand messages only slightly reduces the likelihood that people will share the content. More than 40% said they are more or slightly more likely to send marketing-related messages, while only 5% refuse to share content that contains a clear brand message

Nearly nine out of ten people who receive emails with brand sponsorships said they had no adverse feelings about the brand-related emails. A Sharpe Partners spokesperson told the *eMarketer* website, "It is clear that viral marketing is a low-risk approach."

Sharpe identified a group it called "Brand Fans," who, as *eMarketer* said, "are so viral they are contagious". More than 80% of them feel positively about brand-sponsored content shared at least once a week, and 35% share daily. Brand Fans are also the group most likely to share with 10 or more people.

The study of 1,017 US adults found that the most likely person to share content - and share it widely - is a woman in her late 30s to early 40s; 64% of the female respondents share content at least once a week versus 58% of the males.

The study also found that education is only a slight influence, with 64% of those without a college degree sharing weekly versus 61% with a college degree. Marital status, the presence of children, and household income did not prove to be factors, nor did the length of time someone has been using the Internet.

"We knew a lot of people were sharing content, but even we didn't expect it to be so pervasive," said Kathy Sharpe, Sharpe Partners' CEO. "But, the real challenge for interactive marketing firms is developing content that these people will want to consistently share with a wide, yet focused circle of acquaintances." "Humor is clearly the golden child of viral marketing, but it is also very subjective," cautioned Sharpe. "That is why we recommend a viral conduit that allows the target to define the humor, rather than presuming that we always know what the audience will find funny." "We also discovered that those who share content more frequently are less inclined to view brand affiliation as a negative, so the key becomes targeting those individuals."

REALLOCATING BUDGETS

There are indications that business is already onto the viral marketing trend. A survey of top business executives recently published by

Blackfriars Communications forecasts a 9% drop in the portion of marketing budgets allocated to traditional advertising in comparison with 2005, with most of the shift going to new media and, in particular, viral marketing.

The report, "Marketing 2006: 2006's Timid Start," cites dissatisfaction with existing marketing returns as a major cause for their reallocating budgets away from traditional techniques like advertising to approaches such as word of mouth, buzz marketing, and viral marketing.

Those non-traditional approaches are expected to see a further increase in their share of overall marketing budgets over previous years. According to Blackfriars, non-traditional spending nearly doubled from 8% of overall marketing budgets in Q4 2004 to 14.5% in Q4 2005.

The report projects a 13% increase in overall marketing spending in 2006 – an absolute increase that will offset the large percentage fall in traditional ad budgets, but Blackfriars reports the shift from traditional to non-traditional approaches is still ominous for old-school advertisers.

Blackfriars' 2006 report marks the first time non-traditional approaches have been measured in its annual analysis and forecast, out of recognition of their growing popularity among business executives.

Speaking of the decline in traditional marketing spend over 2005, Blackfriars director Carl Howe said, "We were tracking it for a while, and we watched it descend all year. The first time we saw it we thought it could be statistical variation – but it's not. It's large enough, and consistent enough, to be a major trend showing a big shift away from traditional advertising spending."

Howe also detailed a major difference in reported dissatisfaction between executives whose companies measured marketing results - for example, using click-through and impression rates in Web advertising - versus those who didn't. Executives whose companies measured results showed a 13% dissatisfaction rate, vs. 37% for those who didn't.

Interestingly, Howe pointed out that many of the increasingly popular "non-traditional" methods like buzz and word of mouth are not easily measured; he attributed their growing percentage of overall marketing budgets to an "experimental" attitude among business leadership.

COMMON THREADS FOR SUCCESSFUL VIRALS

Tessa Wegert writes on ClickZ: "Though it seems consumers haven't had their fill of sharable content, that alone doesn't make a successful viral initiative. And although overt brand messages may not deter consumers from sharing at present, the best applications skew more toward product placement than brand sponsorships.

"These initiatives do more than make their way around the Web and tout a logo; they're integrated with the brand and represent it in a creative way. There might not be a true formula for developing a viral piece, but the common thread is brand and product relevance."

The *ClickZ* article mentions a few US campaigns that have been particularly successful in maintaining brand and product relevance. The Virtual Bartender, for example, is a site set up by the Universal Beverage Organization, which is apparently dedicated to getting people to drink more beer (like they need help?). The Virtual Bartenders are "real live" politically incorrect beer wenches who, like Burger King's Subservient Chicken, follow (almost) any command you type into the request box. However, if you ask them to serve you a drink, the only drink they will serve you is a beer.

As Tessa Wegert writes, "It's difficult not to be heartened by the news consumers still spread content online, and equally hard not to be tempted by the apparent simplicity of the approach. Before you attempt to create your own viral offering, however, consider your product and brand. There's a clever angle out there just waiting to be exploited. Miss it and you could ruin the broth."

VIRAL REVOLUTION HITS DOUBLE FIGURES

(PUBLISHED IN 2006)

It's been 10 years since the first viral online marketing campaign. It's a campaign that's still going, and just about everyone who's ever used the Internet has been exposed to the virus.

It was back in July 1996 that the founders of Hotmail, Jack Smith and Sabeer Bhatia, added a signature line to all messages from their email service, alerting anyone who received a Hotmail message that they could come to Hotmail and get a free email account. Nearly 250 million Hotmail members later, that campaign is still running (Editor's note: somewhere along the way after this was written, the line was dropped), and it has been followed by tens of thousands of online campaigns, according to a recent MarketingSherpa report.

However, MarketingSherpa reckons viral marketing has settled into a rut, despite the fact that creativity is the essence of a successful viral campaign. In fact, in its report, based on a survey it conducted of 790 viral marketers, MarketingSherpa said, "it's getting hideously same-old, same-old boring. Which spells trouble, because viral's all about not being boring."

FROM REVOLUTION TO MATURATION

While MarketingSherpa warns of viral marketing being in a rut, marketing consultants Justin Kirby and Paul Marsden argue that the viral revolution has just matured. Kirby and Marsden are editors of the recently published book *Connected Marketing: The Viral, Buzz and Word of Mouth Revolution*, in which 17 experts in the area discuss a range of, as they call it, "scalable, predictable and measurable solutions for driving business growth by stimulating positive brand talk between clients, customers and consumers."

In an excerpt from the book published recently on MarketingProfs. com, Kirby writes that they coined the term "connected marketing" to cover viral, buzz, word of mouth and any kind of marketing (from product research and development through production, to promotion by any means, including traditional advertising) that "creates conversations in target markets that add measurable value to a brand".

Kirby highlights the crisis in advertising effectiveness which has helped fuel the viral revolution. A 2004 study Deutsche Bank in the US consumer packaged goods sector found that only 18% of TV advertising campaigns generate a positive return on annual investment, while the *Harvard Business Review* reported that for every dollar invested into traditional advertising for consumer packaged goods, the short-term return on investment is just 54 cents. The *HBR* survey also quoted a study that showed that an astonishing 84% of B2B marketing campaigns actually result in a *fall* in market share and brand equity.

One response to the problem has been reducing media costs by using free media – word of mouth connections - to deliver marketing messages. As well as being free, word of mouth connections are influential: Kirby cites a 2004 UK survey by consultants CIA:MediaEdge which found that 76% cite word of mouth as their main influence on their purchasing decisions, compared with only 15% for traditional advertising.

Kirby writes, "Many brands are now finally realizing that the most powerful selling of products and ideas takes place not marketer to consumer but consumer to consumer."

Although most viral marketing campaigns look as organized as a men's weekend away, Kirby argues that managing successful viral activities "is possible through an organized series of decisions and approaches; it's not a hit-or-miss quest for that one groundbreaking idea."

CIRCLES OF INFLUENCE

Integrating the viral, buzz, and word of mouth techniques with the whole mix of marketing - advertising, direct marketing, sales promotion, PR, etc. – is essential for successful campaigns.

According to Kirby, successful integration starts with listening to what customers are already saying about your brand and identifying and involving your most influential customers.

"With connected marketing," Kirby writes, "the aim is to profile and recruit customers who represent the 10% of society that helps influence the majority of all purchasing decisions. These influencers... are not necessarily the customers who spend the most money with you, but they are the most important people you can reach, because they are the ones who can help amplify and accelerate positive word mouth about your brand."

He recommends developing ongoing, two-way relationships with influencers, doing things such as letting them trial your product before it is released, or getting them involved back in the research and development process. This will help increase the chance of success even if the product or service itself isn't particularly innovative.

One thing Kirby warns against is jumping into bed with marketing specialists who claim they can monitor blogs, forums etc. to gather data on mentions of your brand. He points out there are plenty of free tools available that you can use to monitor your brand 'buzz'.

MEASURING THE FULL IMPACT

Kirby wisely recommends that traditional, cost-per-thousand-eyeballs types of measures should not be relied on to measure the success of viral campaigns. He also argues against relying on recommendation rates as the only measurement. However, he doesn't offer an iron-clad solution for measurement techniques – he simply says that further research is needed to help marketers be more scientific about success measures.

The MarketingSherpa report, while it also doesn't solve the measurement problem, includes some useful tips for successful viral campaigns in today's social networking-driven online environment. Suggestions include:

- building a campaign specifically with a networking site like MySpace in mind;
- adding secrets or codes your campaign that only insiders of your demographic might notice;
- get celebrities involved, even if it is a parody or imitation of a celebrity;
- brand your creative in such a way that cutting your offer or logo out of it is next to impossible; and
- always add hotlinks such as 'forward to a friend', 'add a permanent link' or 'add to my favourites'.

SOCIAL MEDIA - THE OPPORTUNITY IN THE GLOOM

(PUBLISHED IN 2009)

No prizes for guessing that this year is shaping up to be an ugly time for marketing and PR. Australia may technically avoid a recession, but everyone is behaving like we're in the middle of one, with budgets being slashed and projects put on hold across nearly all industries.

There is a lot of evidence showing that this is the time to increase marketing spend, not reduce it. As the *Harvard Business Review* stated earlier this year: "It is well documented that brands that increase (marketing) during a recession, when competitors are cutting back, can improve market share and return on investment at lower cost than during good economic times."

OK, hands up, all those whose marketing budgets have been boosted for the coming year. Um, yes, I thought so – the reality is that the standard business reaction is to ignore the above-mentioned evidence, scream 'Fire!' and head for the exits when business indicators head south.

LOW COST IMPACT

Well the good news is that tough times present a great opportunity for marketers and PR executives to provide cost-effective solutions and, if overall spend is down, at least you can gain budget spend from the evil empire, aka the advertising guys.

Alan Parker, a digital PR executive whose most recent Australian role was head of technology and digital for Burston-Marsteller Australia, says that "the rise of digital and social media bring plenty of new PR and marketing opportunities, all at a much lower cost than big TV ad campaigns."

He says that the growing influence of the Web on business is leading more and more marketers and PR executives into the digital sphere, although he argues that, particularly in Australia, "The C-suite are still the people you need to convince" in order to win budget approval for digital projects.

A recent study among chief marketing officers (CMOs) by the Epsilon research group revealed that although 65% of CMOs acknowledged that global and local recessions would dent their budgets, digital marketing was seen as a bright spot. Social computing (word of mouth, social networking sites, viral advertising, blogging, etc.) was the most popular emerging channel, with 42% of marketing executives expressing interest in adding it to their marketing mix over the next year.

As US Web 2.0 PR expert Todd Defren writes, "The agencies that survive the coming crapfest will be actively helping their own clients realize the benefits of socializing through the recession."

However, Alan Parker warns that it's not as easy as just "whacking out a press release online". Although he acknowledges that there is still a place for the press release, the humble PR release has evolved, with the ability to add video (or even consist only of video), or a product demonstration with tags and links to more information.

In the digital arena, PR companies need to change their orientation to aim at markets, not clients, he says. "You need to match your client's needs with what your audience is looking for."

Before starting to talk to customers online, it's important to "open your ears and hear what people are saying about your company," Parker says. Some simple ways to do this are to search for mentions of your company on Technorati and setting up Google Alerts.

There is no shortcut to getting this information; it is something that will take time and continual effort, "It requires dedication, it requires effort," according to Parker.

Once you have listened and learned, it is then time to start the conversation, through things such as corporate blogging, engaging with bloggers in your topic area and using networking devices such as Twitter and Facebook.

UNTAPPED AREA

PR agencies are still hesitant about using social media tools such as blogging. Recent US and British research highlighted on the Deep Jive Interests blog, showed that while most PR executives believe blogging is effective for sharing information quickly and broadly (UK 70%, US 80%), and has a role in influencing public opinion and decision making (UK 60%, US 70%), most PR companies don't have a blogging policy (UK 82%, US 88%), and few of them blog for their own company or clients (UK36%, US 37%).

UK-based Copywriter and blogger Matt Ambrose writes that, "Many PR firms are still nervous about jumping into blogging headfirst, preferring to wait for others to test the water and then watch to see if they sink or swim.

"PR's methodology shares some characteristics with that of the advertising world. Their tactics for influencing mindsets have generally been a one-way, top-down approach. In a world where everybody now has the means to share their opinions and experiences with millions of online consumers this approach now seems antiquated. PR, like advertising, needs to learn how to listen to what people are saying and to be able to have their client's voice."

Ambrose argues that "Successful PR in the online world is about more than just generating buzz with a virtual store in Second Life. To get info-hungry consumers to listen to your client's message you now need to trade in the currency of transparency and value. By utilising podcasts, wikis, blogs and RSS feeds, PR agencies have the means to communicate in a more open format, and in a manner which will allow their client's voice to be heard in the online conversation."

Todd Defren, meanwhile, points out that, ironically "the distinct lack of editorial professionalism in the blogosphere" has "forced PR agencies to increase the genuine professionalism of their outreach – indiscriminate e-mail blasts are on the wane."

He warns that "Consumers are ever more resistant to marketing messages that use the old "one-to-many" approach but are often enthused about marketing programs that are useful, empowering and inclusive." Meanwhile, "Web 2.0 technologies have made participation more fun, accessible, instantaneous, trackable."

He says the critical factor to remember about social media is that "it is not only helping brands spread the word; it also helps the brand to shore up support among its current customers. "

LONG-HAUL THINKING

Social media is about building relationships, not jumping in and jumping out of conversations. "Don't think campaign – think long term for social media initiatives," Alan Parker says.

He adds that because social media is still in its infancy in Australia, there is plenty of opportunity for companies who haven't yet climbed on the bandwagon. "We are at the point in Australia where the C-suiters are starting to recognise the importance of the online world."

As technology develops and features such as high-definition video become more widely used in Australia, corporate social media use will really take off. Are you ready for take-off?

BEING THERE: TEND TO YOUR BRAND ONLINE AND REAP THE BENEFITS

(PUBLISHED IN 2009)

Have you seen the classic film *Being There*? The main character Chauncey (played by Peter Sellers) is a mentally-challenged gardener who through a few twists of fate ends up being a respected political adviser and commentator (I can heartily recommend you getting it out on DVD – a funny film with pointed social commentary that still stings today).

Anyway, when he's asked his opinion on world events, Chauncey starts talking about the only thing he knows – gardening – and he slowly and deliberately describes the process of planting seeds, watering them, pulling out weeds, pruning, and harvesting. Everyone who listens to him puts their own spin on what he 'really' means, and he quickly becomes an internationally respected political guru.

Chris Abraham, a guest on our recent podcast, says Chauncey Gardner's gardening analogy is particularly apt for online social media marketing today.

Abraham, online PR specialist and president and COO of online consultants Abraham Harrison, based in Washington and Berlin, argues that despite the right-now, viral nature of the Internet, building a company's brand through social media takes time.

He loves to use real-life analogies. "Building your communications online is like seeding a reef," he says. "You have to hang out in the ecosystem, become part of that ecosystem, occasionally adding things that become part of the reef. And if you're there long enough, the reef builds around you."

Moving from fish to people, Abraham says social media marketing, or as he calls it, 'online conversation marketing', is "like going to a party – you need to understand what the 'lingua franca' is, who your host is, what kind of appropriate gift you should bring, how people talk, and what people expect."

He says companies need to become ambassadors for their brand as they take their marketing online – no broadcasting or shouting in this new environment, just a focus on others and a diplomatic tone.

FIVE DOS AND DON'TS

Another new media marketing consultant, Joseph Jaffe, author of *Join the Conversation* and *Life After the 30 Second Spot*, says there are five key things businesses can do to start participating in online conversation marketing:

- **Listening** – so you can make your contribution to the conversation real, not just hype
- **Responding** – whether approaches are negative or positive
- **Joining in** – making non-partisan contributions to position yourself to be invited to join the conversation
- **Catalysing** – empowering customers to demonstrate your brand on your behalf
- **Starting** – being a conversation conduit and starting a conversation

He also says that companies shouldn't be:

- **Fake** – instead, be transparent in your communications
- **Manipulative** – don't try to fool other participants, but instead be open
- **Controlling** – understand that you can't control everything all of the time
- **Dominating** – the world doesn't operate solely on your terms, allow others room to talk
- **Avoiding** – marketing is no longer a spectator sport, you must be active and participate

Ritu Pant from the Marketing Hackz website, puts it concisely: "Conversation marketing is nothing but a way to gain recognition and create a presence among your potential customers. The only thing that is required in order to carry an effective marketing campaign is the ability to dedicate time and be a part of the community."

"....There is no requirement that you have budget for marketing because it simply requires your time and effort in effectively carrying on a two-way communication. This is one of the reasons why social media has become so powerful in online advertising. If your business doesn't exist on the web, you are pretty much non-existent."

CONVERSATIONS IN TOUGH TIMES

The question on every online marketer's lips is "How will the global financial crisis (GFC) affect e-marketing?" Will it be tougher to get companies to spend money on unproven techniques, or will the cost-effective and measurable nature of e-marketing create a boom amid the gloom?

Chris Abraham has some strong opinions on the issue. "We need to recognize that this is going to be a deep, deep recession - one that's going to last for a long time," he says. "Recessions have major ramifications on how consumers spend their income, how companies formulate their budgets, and, perhaps most importantly, how marketing is viewed. In a recession, marketing is often viewed as an expense...not an investment.

"(In times like the present) decision makers often want to work with 'proven' models that they're familiar with. And these models will often be pushed by their traditional agencies because those agencies provide these services. Of course, (what is) proven may no longer mean effective - but at least it has been done before and for the decision makers, it's best to stick with what is familiar."

Particularly in the current financial environment, he says, "traditional marketing still very much has a primary role.... we can't... dismiss traditional type stuff as being 'so 20th Century'. The end user - the consumer - will be getting the information they seek on products from various sources. "

Abraham says it's important to integrate conversation marketing with traditional marketing techniques. "Social media may not be for every business. Or, more realistically, the emphasis placed upon social media will vary depending on the client's needs and the industry they are in. In practically every case social media will be only part of the equation."

Abraham encourages new media marketers to turn down the hype and turn up the practicality. "Ladies and gentlemen, this is a transformation. An evolution. One that is bringing about substantial change. But the change isn't absolute nor is it complete. People may not want every brand to try to 'engage' them. They may want to just buy something and be left alone.

"We need to stop the shrill 'change or die/nothing will ever be the same' mantras. Yes, change is happening, but we need to remember that we are pioneers and early adopters. Not everyone has a Facebook profile or a Twitter presence and most people don't religiously read blogs."

I've been thinking a bit about Chauncey Gardner, and I reckon that if he was around today he would be a hit on Twitter. The most popular Twitterers are people who dispense timeless common sense that strikes a chord with everyone, rather than those tweeting about the bleeding edge of technology.

As Chauncey says, a patient, long-term approach will help all marketers steer and develop their business successfully through the recession and beyond.

CROSSING THE MEDIA DIVIDE: ARE YOU UP FOR THE CHALLENGE?

(PUBLISHED IN 2009)

There's plenty of talk at the moment about how traditional advertising and marketing is being killed off by the Internet.

For example, Eric Clemons, a professor at a prominent US business school, has written a touchstone piece for the TechCrunch website proclaiming that the social, interactive, search-based nature of the Internet is shattering all forms of advertising – ironically, including Internet advertising itself. His treatise has touched a nerve with the industry, with nearly 700 comments posted at last count either furiously attacking or passionately defending his position.

I think the article raises a lot of good points about the trust issues consumers have with advertising and marketing. But I think it's a stretch to say advertising will disappear. However, it will definitely have to change. Marketers need to take a broader view of how their brand messages are distributed. Integration across different media, including interactive, is the key.

This isn't a brand-new idea. Originally called integrated marketing communications (IMC) or 360-degree marketing, it's now called cross-media or trans-media. Call it what you will, it's a good opportunity to effectively get your message out to a variety of potential customers.

Australian cross-media specialist Christy Dena, director of Universe Creation 101 and a lecturer and researcher at The University of Sydney, spoke to me for a recent podcast, and she says that although the concept of cross-media has been around since the 1980s, the difference now is that the skill set of both the people creating cross-media campaigns and the audiences who consume them has changed.

ELICITING RESPONSES

Back in the beginning, IMC just meant replicating your marketing and advertising campaign consistently across different media – for example, if you used Shane Warne in your TV campaign, you'd use images of Shane Warne (along with the tag line from the TV campaign) in your newspaper, magazine, radio, billboard and DM campaigns.

In this scenario, which Christy calls the design-oriented approach, an eye for blending and consistency were the key capabilities agencies required.

Now, although the Internet hasn't destroyed traditional media campaigns, it's made them considerably more complex. Adding interactivity to the mix means not just broadcasting a message across, it's as Christy Dena says, "Providing content that's appropriate for each audience."

Rather than a wide and shallow broadcasting approach, online audiences are looking for depth of experience, according to Christy. "You're aiming to elicit a response, you're not just telling them something."

The fragmentation of online audiences also means you have to create a multi-layered campaign that has features that appeal to a multitude of audiences.

"You can't design for a massive audience – you can use mass appeal gateways, but you need to offer different levels to different types of people," Christy says.

STEALING CROSS-MEDIA

Many marketers would be aware of the Audi A3 "Art of the Heist" campaign, a textbook example of 21st-century cross-media. Audi of America launched the A3, a premium compact aimed primarily at affluent 25-34 year-old males, with a cross-media campaign revolving around the theft of a prototype A3 from Audi's Park Avenue headquarters in New York.

The three-month "alternate reality game", which started with the online release of CCTV footage showing the car being stolen, blended fact and fiction, encouraging people to help follow the exploits of two recovery experts as they looked for the car and chased the thieves across the company. The search for the car finished at a gaming convention in Los Angeles nearly US$4 million ($5.8 million) later.

And when I say cross-media campaign, I mean cross-media: over the course of the campaign, media used included television, newspapers, outdoor, commuter rail, magazines, websites, blogs, live events, email, podcasts, films, online advertising, direct mail, radio and voicemail.

But what did Audi get for their $5.8 million, I hear you ask? During the first 90 days of the campaign, the Art of the Heist campaign achieved 45 million PR impressions, 2 million visits to AudiUSA.com (double the traffic for the same period the previous year), 500,000 "story participants", 10,000 dealer leads, 4,000 test drives and, finally, 1,025 cars sold within a month of launch and 5,400 in six months.

Engagement in the form of audience participation from the car-hungry gamer community included scores of fan websites that popped up. The agency behind Art of the Heist, McKinney & Silver, won Best in Show at the 2005 MIXX Awards for the campaign.

Christy Dena said, "People who were interested in the format, not just the brand, were attracted to the campaign."

EXTRA EFFORT REQUIRED

The Art of the Heist works on several levels, unlike other more recent examples such as the local Witchery jacket campaign on YouTube, where a winsome blonde who posted a video looking for the stud she met at a cafe who left his jacket behind turned out to be an actress spruiking Witchery's foray into men's fashion. The Audi A3 campaign didn't involve deceit like Witchery's campaign and, as Christy says, "it projected the brand and the image at the same time."

(As an aside, I have to admit that when I watched the Witchery video, one of my first thoughts was, "Nice jacket, I wonder what it costs?")

As the Audi A3 example amply demonstrates, unlike traditional media campaigns, the launch of a cross-media campaign is just the beginning of the process, not the climax.

There are not an enormous number of examples of successful cross-media campaigns to point to (although a future blog post will include links to a few of them). And there's a good reason for that – it's bloody hard to do it. It requires a lot of extra effort from the client, not just the agency. More effort in planning, more effort in execution and more effort in analysis – but with potentially far richer rewards.

It demands a new way of thinking and behaving – you need to retain your old knowledge of traditional media, while building on it to include interactive in the mix.

It does have some aspects in common with traditional broadcast media campaigns. You need to build the biggest, widest funnel to attract people, and then allow them to self-select and decide whether they will move further down the funnel.

Even if your target audience is fairly specific, it's not a good idea to restrict participants in cross-media campaigns, according to Christy Dena.

"You can't say only people who will buy this product can come in." More people equals more buzz, and buzz will greatly enhance the success of your campaign. Besides, you may pick up customers you wouldn't have under normal circumstances.

You don't have to go to the lengths Audi of America went to – and in the Australian market, who can afford to? But in this increasingly complex world, embracing cross-media can give your marketing added relevance.

Are you up for the challenge?

THE GREAT JUGGLING ACT

(PUBLISHED IN 2009)

Does it feel like you're cramming more and more into every day, so much so that you end up doing several things at the same time? Well, that's not surprising, because that's exactly what's happening.

Multi-tasking is the norm for consumers today. A recent study on media consumption by Nielsen found that as overall media consumption grows – more than 140 hours watching TV and nearly 30 hours on the Internet per month – more than half of people are doing both at once.

That multi-tasking increasingly involves duelling video, as Internet video consumption has risen 50% year-on-year to more than 3 hours per month, with 83% of that short-form video.

Interestingly, the multi-tasking trend is equally strong across all age groups, not just the province of teenagers. And although the amount of time spent multi-tasking is increasing, according to the Nielsen study it's only 10 minutes per person per month, which I reckon is severely under-reported.

CHRONIC DISTRACTION

Tony Surtees, executive director of iPrime and an Internet pioneer in both Australia and the US, delivers some compelling observations about multi-tasking in our recent podcast. He cites a Yahoo study which concluded that if you took all the pieces of media an average person is exposed to each day and laid them end-to-end, they would add up to 43 hours per day. That's a lot more than 10 minutes per month!

"People move between different modes of media consumption," according to Tony. "We're all multi-tasking – our kids multitask more out of pleasure than necessity. It's quite natural for teens to have four online chats open simultaneously, while they're listening to music, doing homework, etc.

"Does this mean we're going to end up with chronic attention deficit disorder-laden individuals? Not necessarily," he says, although he points out there are some risks of excess distraction for heavy multi-taskers.

A recent study by a group of researchers from Stanford University found that people who chronically engage in media-multitasking "have more trouble ignoring distractions, keeping irrelevant memories from interfering in their present task, and switching from one task to another, mostly because they can't help thinking about the task they're not doing," according to lead researcher Eyal Ophir. However, as he points out, heavy multi-taskers are also more quick to respond to events in their environment.

For marketers, evidence is mounting that these heavy media multi-taskers are, as Tony Surtees says, "a growing audience of active and engaged consumers that can be targeted more effectively."

He points to recent research from the European Interactive Advertising Association that indicates that media multi-taskers are heavy communicators online, particularly via social networks. They are also more inclined to take in information from brand websites, price comparison websites and customer website reviews when researching products and services, and they are 33% more likely to actively change their mind about a brand than non multi-taskers. Importantly, media multi-taskers buy more things online than non multi-taskers and spend 26% more on the items they buy.

PIECES OF INFORMATION

So if media multi-taskers are increasingly dividing their attention between different information sources, they have less focused attention on that TV spot, or that magazine ad or that banner ad, because they're too busy responding to that Facebook chat pop-up that's grabbed their attention.

Tony Surtees has labelled the phenomenon of getting product information and recommendations from a variety of sources, not just advertisers, the "confetti economy", and he says the most powerful pieces of confetti are those from our social networks.

"Peer support is important in the confetti economy," he says. "We're going to get influenced by peers, family, friends – we will tend to like what they like. They've effectively pre-tested it for us."

"Social media is increasingly a catalogue of the intentions of an entire community. It reflects what's going on in their heads/life. It's very hard for a marketer to use traditional methods without taking into account all these specks of influence that affect the audience they're trying to attract.

"We need to participate in the conversation of an entire community, and listen carefully to what they have to say – the feedback they provide and the consumer behaviour they're displaying."

FROM COMMENTS TO BOYCOTTS

You also need to be careful not to piss them off. New Australian research indicates that nearly 25% of online consumers are prepared to boycott an organisation if they read a negative comment on social networking sites. The survey, conducted by StollzNow Research for RightNow Tecnologies, found that negative comments made on social networking sites by consumers about their customer experience with certain companies were enough to dissuade potential customers from conducting business with them.

It will probably come as no surprise that when respondents were asked which industry generated the most comments, telecommunications topped the list – and 71% of those comments were negative mentions. Government agencies came next at 63% negative, while travel and leisure companies were the most likely to earn positive comments at 73%.

There are a growing number of examples of consumer backlash via social networks. In Australia, a campaign against the Cotton On clothing company's slogans on t-shirts for babies (in particular, the line "They shake me") started via the mamamia blog in August.

Meanwhile, back in July, a country musician in the US angered at being ignored when he complained about the treatment of his musical equipment by airline baggage handlers posted the ode "United Breaks Guitars" on YouTube. It now has more than 5 million views and more than 22,000 comments, and it looks like no amount of money offered to the singer (after the video went viral they offered to pay $3,000 in restitution, which he told them to donate to charity) can overcome the effect of the incident.

But it's not just about dragging companies down. More than half of respondents in the RightNow study indicated that if they did leave negative comments about an organisation on a social networking site, they would invite that organisation to respond in an attempt to help resolve their product or service-related issue.

As Tony says, "If you don't do the right thing by your community, it has the means to address it. You can't delegate this to your PR company or outsource it – you need to engage with your community."

MULTI-TASKING MEANS MULTIPLE MEDIA

Marketers relying on their traditional media experience can take comfort in the fact that all this multi-tasking means there is future for the traditional media models in the digital age. According to Tony Surtees, "Consumers are taking information in multiple forms and simultaneously. Traditional media will survive as long as they adapt – they will blend into a difference architecture of engagement. There will be innovative combinations of messaging – online and mobile, mobile and TV, online and print, etc.

"You'll still have hits – embellished and reinforced by the collective efforts of communities. Reinforce and build on it – focus on content, not product."

His advice to marketers? "Figure out what your consumer wants. Realise your consumers move in groups – market opportunities happen much faster than you realise – time frames become shorter and shorter – integration of an idea/concept into a community."

RULES OF ENGAGEMENT FOR BLOGGERS

(PUBLISHED IN 2009)

They're unpredictable, opinionated, and, particularly in Australia, they're doing it part-time. So why should companies care about engaging with bloggers?

Jason Preston, a US-based social media strategist, writes that more companies these days are turning from blogs to focus on social networking sites. "If you can generate good word of mouth and drive sales from efforts in sites like Facebook, LinkedIn, or MySpace, why bother to court the hard-to-reach and often hard-to-impress blogerati?

"Here's why: because they're hard to reach, and hard to impress, and everybody knows it. These bloggers have spent time building up a brand, and that carries value when they talk about your products or your messages.

"Harnessing this trust, this existing relationship, is why it still matters to work with bloggers who have a name and a following, instead of simply trusting in the effectiveness of blind, stranger-to-stranger word-of-digital-mouth marketing."

Scott Rhodie, the head of social media agency House Party, agrees, saying companies should take bloggers seriously, because they have increasing influence with customers and potential customers.

Scott, who was interviewed for our recent podcast, says bloggers form part of what he calls "the influencer sphere" that consumers rely on when making buying decisions in a post-advertising world. "People look for them," he says." They are independent of product. And while not all of them discuss products and services, some are recommenders."

When selecting which bloggers to approach, it's not just about reach. As Scott points out, a blogger with 10,000 readers may be more authoritative than one with 100,000 followers, and if they blog on a topic closely related to your business, there are powerful targeting benefits.

He says the first thing you need to determine is whether blogger engagement will work for your company. "You need to ask yourself: Am I reaching the right audience? Blogger engagement isn't right for everyone."

One way to work out whether there's value in blogger engagement for your company is to search for terms important to your business and your customers, and see what blogs turn up on the first page of results.

Steve Broback writes on the Blog Business Summit blog: "The best bet is to find approachable bloggers with the right topical alignment.... if you are topically aligned to a significant degree, even a relatively popular blogger can find your message of interest.... find bloggers who are writing about things your customers are interested in, and have aligned posts that are prominent in search.

"Significant and growing numbers of shoppers begin their buying process in a search engine. Anyone with a retail site can attest to the fact that their server logs show the bulk of their traffic is coming from search. Blog posts are featured prominently in results your customers are finding, and these are the bloggers to engage."

So once you've determined which bloggers you want to engage, how do you work out the best way to approach them? Scott Rhodie says, "Do your research. Find out who you're talking to, research their interests, speak to them in their own language. You should never cold pitch." Some experts recommend reading at least six months' worth of a blogger's posts before getting in contact with them.

The global head of WPP, Sir Martin Sorrell, says "You certainly can't spin your way to a blogger's heart. Respect and engagement are essential. Handing over product to bloggers would not be enough. If, however, you invite bloggers in to get their ideas on a brand, you might succeed. Get them involved; give them something of value. The prize for getting it right? The stakeholder becomes a brand loyalist and tells other people."

There's a big difference between blogger engagement and traditional PR, according to Scott. "Blogger engagement requires a different approach that traditional PR and the use of press releases. Each blogger is different, so you need to approach each one individually."

Amid all the bad publicity surrounding companies who offer products, etc. in exchange for positive blog posts, Scott says it's imperative to be honest, upfront and ethical when dealing with bloggers.

"When approaching a blogger who you haven't dealt with before, introduce yourself, who you are, and start the conversation by saying something along the lines of, 'I don't know if you do product reviews, but here's what we're on about.'"

Like with traditional PR, Scott recommends you look for an angle. "You need to take it to them: 'Product A would benefit you because...'

And once you've established contact, you need to make sure you don't leave them hanging. If you said you would send along information or a product to review, do it promptly. "Don't leave them hanging," Scott says.

He also encourages companies to think beyond individual campaigns. "A blogger engagement strategy has to be long term – like social networking, it's about building relationships. You need to allow at least three months before you can expect to see results."

WPP's Sorell says the "willingness to surrender control is essential, because in digital and social media there is an inverse relationship between credibility and control. The more control you keep over the message, the less credible it is. And vice versa."

And in a final word about selecting the right bloggers and developing that relationship, I like what US blog strategist Teresa Valdez Klein has to say. She writes: "Reaching out to bloggers who can influence your niche is smart. Reaching out to bloggers who already like your product and have some reach within your niche is smarter. And reaching out to bloggers who have said negative things about you and trying to improve their experience with your company is the cleverest move of all."

PASS IT ON: HOW DIGITAL TECHNOLOGY AND WORD OF MOUTH ARE A PERFECT MATCH

(PUBLISHED IN 2010)

In the same way that digital technology helped convert the world's oldest profession into the booming global industry of Internet pornography, word of mouth, probably the oldest of all marketing techniques, has emerged as arguably the most powerful selling tool of the 21st century, thanks to the Internet.

Traditional advertising and marketing is in chaos as people turn their back on advertising messages and turn to recommendations from their friends, via social networking tools such as Facebook and Twitter, when deciding what products and brands to buy.

Businesses who can use online tools to cleverly and ethically leverage the power of word of mouth will be able to successfully make the transition from traditional marketing.

Word of mouth marketing online can even be a successful business model in itself. Yelp, a US-based service where people write reviews of restaurants and other local service providers, recently knocked back a US$500 million offer from Google to buy its business.

Justin Kirby, interviewed for our recent podcast, is the founder and CEO of both one of the UK's first digital agencies and Australia's word of mouth marketing firm Yooster, and is a globally recognised expert on word of mouth marketing.

He says that, "Word of mouth is an outcome, not a set of techniques. You need to achieve a commercial goal, otherwise you just have a viral campaign, some creative content that doesn't really get you anywhere. Word of mouth needs to lead to customer engagement, in the form of advocacy."

He cites as an example Budweiser beer's "Wassup" viral campaign, which, when it was launched in the UK, quickly added a catchy term to the cultural zeitgeist, but was accompanied by a drop in sales.

In Australia, Carlton's "Big Ad" and Queensland Tourism's "Best Job in the World" campaign, while wildly successful in terms of downloads, sharing, participation and column inches, were both followed by decreased sales/visits.

WOMA – THE DIGITAL CONTEXT

According to the Word of Mouth Marketing Association (WOMMA), word of mouth marketing is "a pre-existing phenomenon that marketers are only now learning how to harness, amplify and improve. Word of mouth marketing isn't about creating word of mouth – it's learning how to make it work within a marketing objective."

The basic elements are:

- **Educating** people about your products and services
- **Identifying** people most likely to share their opinions
- **Providing** tools that make it easier to share information
- **Studying** how, where, and when opinions are being shared
- **Listening and responding** to supporters, detractors, and neutrals

Justin Kirby prefers to use the term "connected marketing", which is broader than word of mouth, although he says it's important not to get bogged down in definitions. WOMMA, meanwhile, uses word of mouth as the broad term. It has identified more than 10 types of word of mouth marketing, (and claims to be finding more all the time) including:

- **Buzz marketing** - attaching a brand to a popular topic in the news
- **Viral marketing** - entertaining or informative messages that get passed along exponentially
- **Community marketing** - forming or supporting niche communities that share interests about the brand and providing tools, content, and information to support them
- **Grassroots marketing** - organizing and motivating volunteers to engage in personal or local outreach.
- **Evangelist marketing** - cultivating people who love your brand and actively spread the word on your behalf

- **Product seeding** - providing information or samples to influential individuals
- **Influencer marketing** - identifying opinion leaders likely to talk about products and who can influence the opinions of others
- **Cause marketing** - supporting social causes to earn respect and support from people who feel strongly about the cause
- **Conversation creation** - interesting or fun advertising, emails, catch phrases, entertainment, or promotions designed to start word of mouth activity.
- **Brand blogging** - creating (open and transparent) blogs and participating in the blogosphere, sharing information of value that the blog community may talk about
- **Referral programs** - creating tools that enable satisfied customers to refer their friends

Justin Kirby says that WOM is still largely an offline phenomenon, one which has been greatly facilitated by online technology. "Research shows that 95% of all conversations still happen in the 'real world'. However, the digital domain allows us to engage a large number of customer easily," he says.

Marc Guldimann, CSO of Spongecell, points out that the Internet not only makes it easy to share opinions, but it creates an obligation to do it. "Consumers have become increasingly stubborn when it comes to brand engagement as they have been given more tools to research purchase decisions and feel that they are more informed now than ever on products and services," he says.

"They also feel that it's necessary to very publicly share their opinions and personal experiences on the latest products — from iPods to iPhones — to books, celebrities, songs, shows... you name it, someone's got a blog, Facebook comment or online video on it."

NOT A SILVER BULLET

Justin Kirby makes the point that social media is a confusing term, because it is not 'media' in the traditional sense, but is a technology. "Media can exist without an audience — in theory you can have a newspaper, magazine, TV or radio station without an audience, although of course without an audience none of them is financially viable. But with social media technology, you need people in order for it to function in the first place. The real 'media' is people themselves."

He stresses that word of mouth isn't a new marketing technique, but a technologically-enhanced traditional method. "It's simply marketing 1.0, applying Web 2.0 technologies. People are on Facebook so let's speak to them. Why are we not speaking to them in bars and pubs? Because it's easier to track things on Facebook."

And like all marketing methods, traditional or new, word of mouth marketing doesn't work in all instances. Harking back to the four Ps, in particular product, Justin says, "Word of mouth is not a panacea or a silver bullet. If your product isn't very good, word of mouth isn't going to help you."

A HEALTHY MARKET OPPORTUNITY

(PUBLISHED IN 2011)

(Editor's note: This one is a bit meta, as Ray researched and wrote this article as well as being the podcast guest!)

The rise of digital in all its forms – Internet, mobile, social media, online video – has fuelled the shift from selling and marketing products to selling and marketing services, as consumers have replaced manufacturers at the centre of the marketing universe.

Everything from product development to promotion to post-purchase evaluation is today built around understanding and meeting customer needs.

This is abundantly apparent in an area like healthcare. From a product-focused sector based solely on convincing doctors to prescribe medications based on scientific evidence (and a few educational dinners), drugmakers are building portfolios of services aimed at patients and doctors around their brands, helping healthcare professionals tackle issues like patient compliance and health education as direct promotion takes a back seat.

BIG NUMBERS

I discussed the implications of these trends with healthcare digital strategist Ray Welling in this month's podcast. And while the growth of online generally as a medium and a marketing tool has been impressive, the numbers for healthcare are truly staggering.

The Internet has now become the top resource for health questions and concerns – according to the latest Google stats, 65% of people look online for health info, compared to only about 50% who talk to their GP and 37% for friends and relatives. Meanwhile, 75% of people research symptoms online before visiting their doctor, while 70% of them research online after they've been prescribed a medication, but before they start taking it.

Searches on major health conditions such as cancer, diabetes and pain management rose by 25% on average in 2010 – and they were already high compared to searches on other topics. Nearly 80% of health consumers do something offline as a direct result of their online research, from talking to their doctor (49%) to changing their behavior (27%) to trying an alternative treatment (16%).

Mobile searches (from either a smartphone or a tablet), which are already growing quickly (more than doubled their share of searches in

2010 to 15%), are increasing even more quickly for healthcare topics. Branded mobile health queries rose by 1400% from 2008 to 2011. Meanwhile, 48% of consumers want healthcare or medical related content on phones in the next 12 months.

The fact that these numbers keep growing shows that it's working: the Internet helps people make better health choices. This represents a clear opportunity for health marketers to provide credible, trusted information for patients and healthcare professionals.

HEALTH MONITOR IN YOUR POCKET

But websites and social media aren't the only digital opportunities for health marketers. The rise in mobile use (particularly smartphones) is tailor-made for the development of useful applications for maintaining and improving health.

Investments by pharmaceutical companies in smartphone apps (along with social media platforms and wireless devices such as iPads) grew 78% in 2010, according to Ernst & Young's Pharma 3.0 global pharmaceutical report.

Andrew Tolve writes in the eyeforpharma blog: "A properly designed app – be it a medication tracker, a disease calculator, an educational catalogue, or a patient diary – can improve the lives of patients and physicians and thus increase customer collaboration.

"At the same time, a well-designed app in the hands of sales reps and marketers can maximize organizational efficiency and create competitive advantage when it comes to selling drugs."

The iPhone in particular is loaded with sensors that can be adapted in ways that can turn your phone into a blood pressure or blood glucose monitor, a stethoscope or even an ECG machine. There are already some apps available that make use of these capabilities, as well as a growing number of apps that encourage patients to manually input data to monitor their diabetes, fertility and cardiovascular health.

With stats showing that people who use in-home monitoring devices have 50% fewer problems related to their condition, apps are proving to be a simple way to promote better health and save on hospitalization and drug costs.

The key is to be able to provide personalized info simply and in an actionable way. Thomas Goetz, medical editor for *Wired*, calls this approach "information with feeling". The example he uses to explain the concept is the speed meters used in some road construction zones – the ones that say "the speed limit is 40kmh; you are going 58kmh, PLEASE SLOW DOWN".

Goetz says this type of approach, and not the current health management approach of lecturing people about what they should/should not be doing, has a much stronger positive effect on behaviour.

Specific, personalized information, he says, creates a connection through relevance, which leads to choices, and finally to action.

KEEPING COMPLIANT

There's a no-brainer when it comes to opportunities presented by consumer health mobile apps – patient compliance.

As Ray Welling says, "There's quite a wide spectrum on compliance, and it relates to understanding the impact of staying on your medication. Women taking the Pill have 95% compliance, because if you miss taking it, you know exactly what will happen and how quickly.

"At the other end of the scale, compliance rates for chronic conditions such as congestive heart failure, diabetes and glaucoma, where you can't really see the difference day to day, are down around 40% – although in the case of glaucoma, this increases to nearly 60% once you've lost sight in one eye!"

Mobile apps are already making big strides in this area. The format of apps lends itself to compiling information and feeding it back to you in a way that gives you personalized advice on your health or your medical condition. There are scores of apps that have been developed to track weight, diabetes-related info, fertility, etc.

The key, according to Ray, is to "provide an app that is actually useful, not gimmicky, and the less information the patient (or healthcare professional) has to input manually, the better."

THE PROFESSIONAL SIDE

With all these opportunities on the patient side, healthcare marketers, particularly in Australia where direct to consumer advertising is not allowed, shouldn't forget their traditional market: healthcare professionals.

Stats are showing that doctors are among the most likely demographic to own a smartphone: 75% of US physicians were using one last year. And they're using multiple devices; anecdotal evidence points to a very high take-up of tablet devices such as the iPad, as well.

As a result, one of the most popular apps in the iTunes store is the Medscape medical news app, which has more than 700,000 users.

This means healthcare professionals have a high expectation for interactivity in information, which is putting pressure on pharmaceutical companies to supplement their traditional paper-based detail aids with tablet-based animations, videos and interactive clinical data, known as e-detailing.

Ray Welling points out that patient-based web information and apps, as well as e-detailing, are fairly undeveloped areas in the Australian market. "Most of the most-trafficked health websites for Australian audiences are American or European health info portals. There's a real need for credible, Australian-based information on a variety of health conditions. Also, Australian pharma companies have been slow to embrace e-detailing. There's a lot of room for growth in both of these areas."

GETTING TO KNOW YOU: DIGITAL AIDS TO PRODUCT FAMILIARISATION

(PUBLISHED IN 2011)

Who do people trust when it comes to health information? Well it's encouraging to know that people are not relying on Kim Kardashian or Lara Bingle for advice on healthy living.

The recently released Edelman Health Barometer surveyed more than 15,000 people in 12 countries, and when they asked people how credible different types of people were in terms of providing health-related information, celebrities came up last, at only 17 percent.

Not surprisingly, doctors topped the list at 88%, followed closely by pharmacists, nurses and nutritionists/dieticians. What is surprising is that some of the most credible information sources were 'ordinary' people - someone living with a disease or condition, and friends and family members.

I think this is a reflection of the rise of social media and the way it has changed marketing across all categories. There's an increasing emphasis on recommendations from trusted contacts (as opposed to celebrities) and a declining reliance on traditional advertising when making buying and lifestyle decisions.

But in healthcare, healthcare professionals are still the most trusted sources of information, particularly in Australia, where regulations restrict discussions between pharmaceutical companies and consumers.

That creates a number of challenges when it comes to familiarising doctors with new medications that become available. Traditionally, the most common method of getting doctors exposed to new therapies is sending pharmaceutical reps out on the road to call on doctors. But today, digital tools are being used to make rep contacts more effective, and to aid the process in other ways.

Product familiiarisation programs, where doctors are able to prescribe a new therapy to a limited number of patients before the medication is available on the Pharmaceutical Benefits Scheme, are one area where digital solutions are having an impact.

Online program registration provides reps and management with a real-time view of how the program is tracking, while also making it easy for doctors to enroll and providing them with resources about the medication at their fingertips.

A lot has been written about the explosion in medical apps for consumers and healthcare professionals. But when it comes to product familiarisation, the most powerful way that apps can function is as tools that reps can use to provide doctors directly with important information about a new medication, as opposed to producing apps that are distributed to doctors.

Many salespeople have been worried that digital tools will replace them, but the fact is that pharmaceutical sales forces have been declining in size for years, independent of digital developments.

One of the biggest developments in digital healthcare marketing has been the rise of e-detailing, where reps use tablet computers, rather than static, complicated, expensive, hard copy sales kits to discuss the therapeutic action of a new medication.

Bill Drummy, writing in *Medical Marketing & Media*, commented on the explosion in use of iPads (and it is almost exclusively iPads at this stage) by pharmaceutical sales teams in an industry that has traditionally been slow to adopt technology.

"In contrast to all earlier waves that washed over the business landscape, pharma doesn't appear to be following its 'follower' instincts – i.e., waiting to see if the platform proves out before jumping on board," he wrote.

In the US, half of the top 20 pharmaceutical companies are using tablets to aid their product familiarisation programs, which is pretty impressive when you consider that the iPad has only been available for about 18 months and its Android competitors are six months old or less.

Drummy writes that there are five aspects of tablets that have led to its quick uptake: 1) the fact that it's instantly available; 2) you can control it with your fingers to create an involving experience; 3) it's easy to move between different media; 4) GPS and Accelerometer technology can provide location-based information and respond to movment; and 5) its size brings the rep in close with the doctor.

Many pharmaceutical sales forces in Australia are now employing iPads in their presentations to doctors, but I think most companies are still in the 'gee-whiz' phase, where they create a PDF version of their current sales kit and everyone oohs and ahs as the rep sweeps from one page to the next with their fingers.

There is a lot more that can and should be done, such as showing animations of mode of action, playing videos and even using the iPad's camera to conduct live video conversations.

Over the next few years, apps developed for tablet computers will quickly evolve into ways that will help doctors to understand much more fully how new medical treatments can change the lives of patients.

ECOMMERCE

THE ARRIVAL OF ESHOPPINGTOWN

(PUBLISHED CIRCA 1998)

The Web is attracting more of the 'shop 'till you drop' crowd. The problem is, technological hitches and poor service mean many of them are dropping before they reach the checkout.

Have you noticed how some new ideas come into vogue, have a sweeping effect on society, then become blended with traditional ideas before fading out of favour? Like the idea that men's and women's brains are exactly the same and society is to blame for boys wanting to play with trucks and girls wanting to play with dolls.

I grew up during that period when parents were told to buy their sons dolls and their daughters tool kits, when high school girls were encouraged to take up wood turning and high school boys to study home economics, when university housing turned co-educational because we were all just people - albeit people with radically different physiques, as the co-eds discovered.

Thankfully, the next generation of sociologists and biologists discovered that while men and women were equal, their brains contain subtle yet significant differences in their wiring, which meant that boys, as a rule, prefer dealing with things (such as trucks and working with their hands), while girls generally prefer dealing with people (starting with dolls and moving to shopkeepers). Once again, science proved what our grandparents already knew.

What has this got to do with the Web? My point is that the Internet, which started out as mainly a geeky guys' thing, with no pictures and a heavy emphasis on impenetrable numerical codes, has evolved into a colourful, visual, functional medium which has now become mainstream. It now has a broad appeal to women as well as men, thanks largely to two gender-stereotyped developments - chat and online shopping.

Yahoo!'s list of most frequent search terms in April had 'chat' at number three with more than a million hits (admittedly still far behind 'sex' with 2.6 million). 'eBay', meanwhile, the leading auction site, was just nudged out of the top 20 by 'Pamela Anderson'. Okay, maybe we've still got a little way to go before there is true gender equity on the Web, but the balance is shifting.

The development that has convinced me that the Web has become mainstream in Australia was the announcement the other week that Westfield was planning a series of Internet shopping centres to join its collection of traditional Australian and US properties. When Frank Lowy decides it's time to join the online revolution, you know some serious shopping is going to take place.

And shopping is something women want from the Web. According to a recent Businessweek/Harris poll, 31% of online users in the US bought something online in 1998, up from 19% the previous year. This year, nearly 30% of those buying online are expected to be women, up from 20% in 1998, according to Forrester Research. Marketing

112

experts, meanwhile, estimate women make 80% of all consumer buying decisions.

Although more than twice as many men currently buy goods online, women are dominating online commerce in areas such as travel, gifts, clothing, toys, flowers, and cards.

Australians are doing plenty of window shopping online. According to the Australian Top 100 website (www.top100.com.au) the top 20 most popular sites with Australians include the Yellow Pages and the Trading Post, along with sites like Amazon.com.

What's frustrating emerchants is that it's just window-shopping. While hard data is hard to come buy, it's estimated only 2% of visits to ecommerce sites result in actual sales. Of those who try to buy something, between 50% and 65% abandon their shopping trolleys before making it through the electronic checkout and into the virtual parking lot with their purchase.

In Australia, bad ecommerce experiences are turning many users off online shopping. A recent www.consult survey revealed 33% of Australian Internet users oppose online shopping, almost three times the percentage in a similar survey three years ago. As the Web becomes mainstream, user expectations are rising. Most of the current ecommerce solutions just aren't good enough.

Security is still an issue, but it's becoming less and less of a problem as cases of online credit card fraud drop off. A bigger problem is the weighty handling and shipping charges levied on online purchases. Other shoppers jettison their trolleys when faced with registration forms that try to capture personal information to be included in a direct marketing database.

But the biggest stumbling block to women - for that matter, all but the most technophilic online shopper - is that it's just too hard. Logging onto the Internet and visiting an ecommerce site is a lot quicker than getting in your car and driving to the mall. But people forget that part when faced with a bewildering array of forms and steps at the end of the online shopping process - they just remember that handing over cash or a credit card to a live checkout person isn't nearly that complicated.

One of the barriers to be overcome is the inevitable comparisons between shopping in a traditional and a virtual store. A different set of expectations needs to be built up. Ecommerce will never replicate the 'touch and feel' experience of retail shopping. But it can be a lot easier and simpler than shopping in real-time. Reducing the time factor in buying online will be a key to making ecommerce take off. One of the main reasons people shop through their computer is because they don't have the time to make it to the store.

For those business who get it right, there is a rapidly growing pool of customers with money to spend. Visa International predicts the worldwide market for ecommerce will be more than $1.5 trillion a year by 2003, representing a compound annual growth of 66 percent.

But for that figure to be achieved, ecommerce is going to have to get simpler, more intuitive. The electronic shopping experience is by definition different from shopping in person, so there's no point simply trying to imitate the real-life experience. The goal should be to use the unique features of the online medium to create a shopping experience that complements the retail experience.

Until that transparency of ecommerce technology happens, many of the big boys making those huge ecommerce predictions will be crying.

A MEDIUM UNLIKE ANY OTHER

(PUBLISHED IN 1999)

IBM's current ebusiness television campaign does a nice job of selling the benefits of doing business online, even if it doesn't provide any clues as to how ebusiness is actually conducted.

My kids' favourite one is the Dr Seuss-esque search for the elusive socks, while mine is the one where the Internet business manager, asked to justify his budget in terms the board can understand, says, "For every dollar we spend, we'll get two dollars back in increased sales." My reply to that man, to quote from another TV ad (yes, Internet developers still dabble in old technology), where a group of boys at McDonald's try to top each other's gustatory tale, "As if!"

That sort of simplistic, measurable relationship between what you put into a website and the return you get is still a far-off dream. The reality, as anyone with a site knows, is much more complicated. And it is the complicated, technology-focused state of the Web that has to be changed if the Web is ever to become a universally-accessed medium.

While the uptake of Internet use is rising dramatically (nearly 20% of Australian homes were hooked up to the Internet in February this year, according to recent Australian Bureau of Statistics figures), it's not because someone is out there making two dollars for every dollar he spends on his site. And while the Web is rapidly establishing itself as a significant new medium, it's not at the stage where it can become as universal as TV, radio, newspapers and the cinema.

For the Web to be universally used, three things need to happen. It needs to be affordable; it needs to be easy to use; and people need to decide it's worth spending significant amounts of time on it.

The first issue is in the process of being licked. Factors such as AOL's entry into Australia, which has driven down the price of big operators Big Pond and Ozemail, and the appearance of pay-by-the-minute ISPs such as AAPT are bringing the cost of Internet use into line with alternative forms of entertainment. The cost of the hardware needed to get started if you don't already have a PC is still a barrier, but the recent application of the mobile phone principle - make the hardware cheap and make money on access - by retailers such as Grace Bros and Harvey Norman is making Internet access more affordable as well.

115

However, the Web still has a long way to go in terms of being easy to use. There really has been no significant advance in terms of making things easier for the average punter since the development of the original Mosaic Web browser, the tool that evolved into Netscape.

Is it any wonder that Australians' most popular online destinations, as tracked by the Australian top100 site, have consistently been search engines and the Yellow and White Pages? People go onto the Web to find things, and those new to the Internet are bamboozled when faced with the almost limitless choice of places to go. They need help finding things. They need a guide to making the experience simple and understandable. They need ecommerce to be simple and transparent before they will start spending serious money online.

Before the Web can truly blossom into a viable medium it needs to demonstrate to people that it is worth taking time away from other activities and spending it in front of their computer screen, an activity most people associate with work. And while the Internet offers a wide range of options to users - information, entertainment, time management - it is the ability to do business that will prove to be the most important distinguishing feature of the Web.

BUSINESS' KEY ROLE

More than any other medium in history, the development of the Internet is being driven by commerce. Subscription sites, for example, the online equivalent of pay television, appeared very early in the piece. Most medium to large businesses and a rapidly increasing number of small businesses have their own website, their own piece piece of the Web. How many businesses own their own television or radio stations or newspapers?

On the other hand, books, movies, radio and television are all entertainment-driven, while newspapers are mainly information-driven. There is a key component of both entertainment and information in the Internet, but using the Internet to either buy goods and services or make the process of buying things offline more effective is becoming the key role of this new medium.

That makes it ultimately more important for business to get on top of the Web than any other medium - it becomes an integral part of the business, one you have more control over compared to other media.

SIMPLE=SUCCESS

Those businesses who are on top of the Web today are the ones who make the Web a more inviting experience by making it simple and capitalising on its unique features to make it worth spending time online.

The sultans of simplicity include technically unambitious sites such as Yahoo! Amazon and iQVC, the online arm of the US' most popular home shopping television channel. And ironically, many of these sites are relying more and more heavily on offline marketing to drive people to their site.

Established brands, in particular, are starting to rely on this technique, quoting their website address on all their communication pieces, or in the case of the Yellow Pages, taking the extreme step of running ads with nothing but their website address as the message.

That's not to say that online marketing is ineffective. The Web is a great place to learn about things before you buy them. As Larry Magid writes in Upside, "Marketing on the Web is, in some ways, easier than on TV, radio or in print. With an offline ad, the best you're likely to accomplish is to whet the potential customer's appetite. If they're interested, it's up to them to call you on the phone, or, even harder, visit your place of business. And you still don't have a sale. Online you can get their attention and, once you do, you can provide an incredibly rich amount of information about your product and services and, in some cases, turn the window shopper into an immediate buyer."

The only problem with Magid's argument is that it is the information-rich preaching to the information-rich. While everything he says is true, most people are still squeamish about conducting business online, for reasons of privacy, security or a lack of trust. And those are the ones who are already online.

The increase in offline marketing will go a long way toward making the Web a ubiquitous medium. There is still a huge untapped market that needs to be convinced that it's worth spending the money and taking the time to get online. Maybe the bloke on the IBM ad should have said, "For every dollar we spend marketing our website offline, we'll make two dollars online."

AMATEUR HOUR MEANS ECOMMERCE SUCCESS

(PUBLISHED IN 2000)

As more and more shoppers and businesses are getting together online, a lucrative business opportunity has arisen in analysing and predicting consumer behaviour in this new buying environment.

Corporations in the US are paying US$8,400 a month or more for advice from companies such as the three-year-old Internet interactive marketing software and services consultancy Personify. It's being paid that kind of money to monitor and analyse the habits of online customers and extract behaviour patterns that can be used to predict whether and why a customer will buy something off a website.

Personify gathers every piece of information generated by visitors, every log file and each transaction, and compiles a customer profile for everyone who has interacted with the site. For those companies willing to spend the money, interactive marketing software outfits provide quite sophisticated analysis of online customer behaviour.

But while the technique is complicated, the conclusions arising from these companies' analyses can be quite simple. One of the key behaviours of online shoppers that has emerged, one which has important implications for marketers, can be summed up in five short words: Experts shop, but amateurs buy.

In contrast to conventional wisdom, the high-income "experts" and aficionados that most businesses are catering to when they build their multi-purpose, all-singing-and-dancing websites are, to put it politely, nothing more than window shoppers. They visit, suck a site dry for information and buy their goods elsewhere, if at all.

Amateurs, meanwhile, visitors whose initial knowledge base is quite low, are the ones who are spending the money online. They're the group at whom ebusinesses should be aiming their marketing spend.

For example, when the US website Virtual Vineyards set up its online bottle shop, it figured that wine snobs would form the core of its online customers. As a result, most of its marketing budget for the site went on ads in specialist gourmet magazines and websites.

When the site failed to exploit its potential, the company enlisted Personify to analyse Virtual Vineyards' web traffic. It turned out that the snobs made up nearly 30% of visitors, but only accounted for 1% of sales on the site. Meanwhile, at the other end of the food chain, "amateur wine drinkers" made up only 8% site visitors but clocked up 82% of sales.

It seems not only the experts were taking advantage of the world's greatest research tool. While the online checkout counter was popular, the most popular stop on the site for the amateurs was the basic glossary of terms. Thousands of people who didn't want to look ignorant in front of their bottle shop attendant used the anonymity of the website to build their knowledge - and then they realised they could completely circumvent the bottle shop by buying online.

This pattern has carried over to other sites. A golfing supplier, for example, found its online promotion budget was better spent on ads placed at a general sports and leisure site than a site for golf nuts.

CATERING TO THE AMATEURS

Because the average web user profile is changing to embrace more of these amateurs, their propensity to buy, in contrast to their tight-fisted expert counterparts, is likely to lead to an explosion in ecommerce.

But getting them to open their wallets is not as easy as just setting up an ecommerce operation and letting them in. They have high expectations about ecommerce - they want it to be simple and convenient.

These amateurs are the unreasonable types who expect a web page to download in less than 10 seconds and who abandon their online shopping trolley at the first sign of complications. Research is showing that these amateurs are the greatest untapped source of Web revenue around.

A recent study by Zona Research (called "The Need for Speed") found that one-third of online shoppers waiting for web pages to down-load will bail out after only eight seconds. This could cost web merchants as much as $4.35 billion in lost ecommerce sales this year.

New Web users' demands for faster, simpler service will drive improvements in ecommerce technology. And when ecommerce truly becomes easy and convenient, the multi-billion dollar predictions of ecommerce revenue that up to now have seemed to be plucked out of thin air may not look so silly after all.

REVENGE OF THE OLD ECONOMY: WHO'S SELLING NOW?

(PUBLISHED IN 2000)

Ah, the speed at which the Internet develops. It seems like only yesterday – in fact, it was only yesterday – that anyone with a PC and an idea for an online business was fêted by venture capitalists and stock markets alike. Amazon.com was worth more than just about every other bookseller in the world combined, and an Internet company became the senior partner in a multi-billion-dollar merger with the world's largest media conglomerate. Traditional businesses were ridiculed for not getting "it" (the Web) and senior executives from such august consulting firms as McKinsey and Andersen Consulting were bailing out to become online grocery and toy retailers.

The "old" business world was slow to respond to the challenge. It was less than 18 months ago that David Pottruck, president of Charles Schwab, coined the term "clicks and mortar" to describe the nascent attempts by traditional brick-and-mortar businesses to enter the ring with the online start-ups.

Since then, Pottruck has written and published a best-selling book about how Schwab took the fight to E*trade and became a clicks and mortar heavyweight, and his buzzword has been beaten to death through overuse. More significantly, thanks to April's stock correction, pure-play online businesses from Boo.com to Amazon (anyone in Australia remember TheSpot?) have been dying off or had their stock devalued by more than 75 per cent, while traditional offline businesses have been storming in to fill the gap left by the dotcoms, in many cases buying them out for peanuts and using their technology.

TRADITIONAL RETAILERS ASCENDANT

Lands' End sells more clothing online than any other company – an estimated $136 million (10% of total revenue) this year and double last year's take. The top website for office supplies globally is 13-year-old Office Depot, with estimated sales of more than $300 million this year, analysts say, of which $30 million will be profit.

The transition to clicks and mortar began before the tech stock crash. During the final quarter of 1999, nearly half of the 50 most visited sites were connected with traditional offline companies (Media Metrix figures). Dell and Intel, who now get more than half their revenue from the Internet, last year each had total revenue of more than $25 billion in online sales. US carrier Southwest Airlines last year sold $725 million worth of tickets online, or 16% of its revenue – more than the combined revenue of online travel agents Expedia and Travelocity combined (PhoCusWright figures).

Brick-and-mortar retailers with strong online sales strategies will grab a dominant share of all online consumer spending by 2002, according to a report by Giga Information Group. The report predicted "multi-channel retailers" – companies that sell products in stores and over the Web – will grow their share of the ecommerce market to two-thirds, or $92 billion, by 2002, compared to one-third of the market today.

POST-WEB RETAILING

The successful online retailer of the future will be the one who blends offline infrastructure with online expertise. The World Wide Web didn't exist 10 years ago, and now we're already hearing about "post-Web retailing". US online research firm Forrester Research recently released a report describing post-Web retailers, those sellers in the new millennium that will understand their customers well enough to be able to anticipate their needs.

They will centralise all available information about shoppers, gleaned from online and bricks and mortar activity, and will be able to identify their most profitable customer segments. They will then identify and adopt new product categories that will appeal to those profitable customers. True post-Web retailers will move beyond personalisation, the current Holy Grail of online marketers, to anticipatory selling.

The store of the future will have both an online and physical component, and is likely to look like a cross between a department store and a category killer specialist outlet.

Unlike pure-play Internet companies – which have quickly burned through cash by essentially giving products away and running costly brand establishment campaigns in an effort to get business – traditional firms already have the expensive things in place. Customers, brands and necessary buildings are built. Adding the Internet is just the gravy of a new channel.

So were online-only businesses merely cannon fodder for big business, sent in to soften up the playing field so they could run on and make the serious money? I think that's ascribing too much intelligence and foresight to big business.

It's more a case that despite the unique nature of the Internet as a business medium, traditional cycles of business development apply. Someone gets a good idea at the right time, puts a lot of blood sweat and tears into making it work, and then big business comes in, buys the person or the idea, and uses its marketing and distribution clout to turn it into a real money-earner.

The great thing for consumers is that the no matter who they are buying from online, customers have more control over the sales process than ever before.

SELL-OUT AT THE DOT.COM CORRAL

(PUBLISHED IN 2000)

The Internet revolution has often been compared to the settling of the American West in the 19th century. If that's the case, then we must be rapidly approaching Manifest Destiny: the West Coast is within view, while to the east towns are popping up wherever there's a water hole. The online equivalent of the transcontinental railway – broadband – is growing from two disparate coasts – cable and ADSL – toward a meeting point that will mean greater communication and more fertile fields to till... anyway, you get the picture.

Most of these analogies consider the pioneers of the Internet to be the folk in covered wagons, who travelled into the unknown blazing a trail for truth, justice and the capitalistic way. But I don't think they've got the analogy quite right. I think the true Internet pioneers – the early adopters, the pure-play, dot.com businesses – are, in fact, more like the original inhabitants, the native Americans.

Think about it: Pure-play businesses were the first ones to occupy the land, they were more interested in engaging with their territory and exploring it than making money out of it, and when the "civilised" interlopers came, they slaughtered them – for a time.

Eventually, the more populous, better-organised and better-funded capitalistic European hordes won out, and the Indians were either subdued or exterminated, their land taken over in the name of progress, to extend the reach of "civilisation." The only Indians left today survive by selling trinkets to the white folk.

Take a look at the Australian online marketplace. Only 12 months ago, the dot.coms were slinging arrows and tomahawks at the white (-shirted) guys and many of them were finding their marks. Traditional businesses were terrified of being murdered in their beds by stealthful online competitors.

A year later, The Spot, ozbuy.com.au, buy.com.au and even dstore lie scalped on the battlefield, while their old-world, "civilised" competitors such as David Jones and Harris Scarfe have a fistful of new hairpieces. The only Australian dot.com of any significance left is Wishlist, which has survived largely through alliances with offline suppliers, building ecommerce sites for companies like Country Road and fulfilment deals with BP.

The trail of carnage has been the same all over the world. In the US, even the greatest Internet warrior of them all, Jeff Bezos of Amazon.com, is running his fingers through his hair and looking behind him.

RISE OF THE CLICKS & MORTAR BRIGADE

The 2000 holiday sales period is proving to be a showcase for the new order. After a brief flirtation with pure-play e-tailing, consumers around the world are returning to their trusted favourites this year – online.

All those abandoned shopping carts, late deliveries and sold-out lines from 1999 are coming back to haunt online merchants. Most of the growth in online sales this Christmas is predicted to come from "new" entrants – traditional companies moving online.

There are some very practical reasons behind the change – the offline back-up provided by traditional businesses. A recent Deloitte & Touche/ NRF survey found that 31% of brick and click marketers allow buyers to return their online purchase to a store location, while 44% offer the choice of picking up and returning to the store.

Meanwhile, a PriceWaterhouseCoopers survey found that 9 out of 10 respondents preferred physical stores to online stores for returns and exchanges of goods.

Clicks and mortar businesses are now dominant in several categories of online sales, commanding 97% of event tickets, 85% of computer hardware/software and 81% of apparel and sporting goods online, while books and music/video sales are still dominated by pure-play businesses (74% and 85%, respectively). (eMarketer figures)

WAITING FOR COLES/MYER?

KPMG is predicting that by 2005, 18% of all shopping will be done online, with Australian consumers spending $50 billion per year. But that rate is well behind most other first-world countries, despite the fact that Australia's rate of Internet usage is among the highest in the world.

Why is Australia slow on the e-tail uptake? North America's catalogue culture is one explanation, but a more likely one is what Niki Scovak from internet.com calls "the oligopolistic nature of Australia." The mere existence of Coles/Myer, Woolworths and David Jones is, she writes, "a huge barrier to pure dot.com retailers gaining critical mass. Perhaps the inevitability of an incumbent victory is the real factor slowing Australia's adoption of e-commerce."

Just like in the westerns, the deck is stacked against the Indians.

MAKING THE WHOLE WEB WORK

(PUBLISHED IN 2001)

When the dot com collapse continued for six months after the tech wreck last April, people said the worst was over. They were wrong. Of the 210 Internet-related businesses in the US that officially curled their toes up during 2000 (not including shotgun weddings and the death of hundreds of small e-businesses too small to appear on the radar), 120 of them failed during the last quarter of the year, the majority of those just before or just after Christmas.

In the aftermath, most analysts have described the collapse using analogies about traditional big business eventually triumphing over the little guy. "The market provided seeming validation for the idea that small companies following the new rules could indeed change everything and topple Goliaths. This created an attitude of invincibility that might have led companies to risk too much," Saul Hansell wrote recently in *The New York Times*.

"If anything is unarguable amid all the dashed hopes and wagging fingers, it is that it takes longer to change the world, or even build a business, than it does to make a pretty website."

An editorial earlier this year in *Fast Company*, a magazine that has always been particularly bullish on the new economy, admitted that "certain assumptions about the Internet economy have been shattered: the short-lived notion that there would be a separate business sector devoted exclusively to dot coms; the assumption that whenever a startup met a big, established company, the question wasn't whether the startup would win, but when."

But they also acknowledge that the Internet has had a lasting effect on the way the world does business. According to Saul Hansell, "The simple truth seen by the web pioneers remains: this medium can connect more people to information and to one another faster and cheaper than any before it. And amid the prominent failures, there have been remarkable successes. So now it's possible to start asking what has been learned from this experience."

After mourning the death of the dot com sector, *Fast Company* went on to proclaim, "Digital technologies remain a powerful force for strategic transformation. Thanks to what's happened over the past five years, companies and their executives have embraced – and will continue to embrace – powerful new ideas about what it really takes to develop a winning strategy, how to build a creative and productive organization, and what it means to succeed."

MAKING BOTH HALVES WORK

I think much of the slump in pureplay ebusinesses can be attributed to the fact that people are still working out how to get an adequate business return from the Internet. As it stands at the moment, the Web only half works. It holds enormous promise, but many businesses tried to go too far, too fast. Pureplay and clicks and mortar businesses that have survived have done so because they learned a few lessons along the way:

Respect the technology – but don't bow down to it: Lots of dot coms poured too much money into developing complicated technology that either no one quite understood how it worked, or was too labour-intensive for too-little return. Early Web pioneers like CDNow and Amazon.com built their business by relying less on whiz-bang technology and more on having a first-class call centre and distribution network.

Most Australian businesses still don't get this. Visit any of the telecommunication, tourism or car sites in Australia. Not only is it hard to work out how to conduct business with them online, the customer service is simply not there backing it up.

Bundlers buy: The main reason why clicks and mortar businesses will ultimately win out is summed up by the latest e-marketing buzzword: "channel bundling". It refers to the fact that an increasing amount of people researches their purchases online, but make the actual purchase in person. Nearly 50% of the most sophisticated shoppers found items they wanted online and bought them in a store during 2000, according to a study by the Institute for the Future and marketing consultants Peppers and Rogers Group – twice as many as in 1999.

Channel-bundling consumers have more power over the total shopping process, according to Peppers and Rogers, controlling everything from the price they pay to how and when goods and services are delivered, and how they're billed.

"The 'store' is no longer at the mall, on the Web, or in a catalogue – it's wherever the customer is when he chooses to shop, browse and buy," the study concluded.

Always have a human available: Amazon.com is still ahead of Barnesandnoble.com – and most Australian booksellers online – because despite the fact that it has never had any face-to-face contact with its customers, dealing with Amazon feels more like you're dealing with a real human being than any of its competitors. From the no-nonsense design of its site to the friendly, intimate tone of its newsletters and the language employed in its correspondence, you know the place is run by real people who love books.

While I'm on the subject of humans, another essential ingredient for a successful web business is to have real people answering any correspondence that comes in – and to do it promptly.

The old economy and the new economy are converging into something that will just be called the economy. And just like in the old days before the Web, businesses that get the basics right and focus on customer service will be the ones that survive.

WHAT GERRY HARVEY AND MYER CAN TEACH US ABOUT SHOPPING ONLINE

(PUBLISHED IN 2010)

It's amusing to see the country's biggest retailers on TV pleading their case for a GST on imports and talking about Australian workers and jobs. After decades of sourcing the lowest possible manufacturing prices from India, China, and Vietnam, we have to defend Australian jobs by paying GST on our overseas Internet purchases under $1000.

Now while these observations are a pretty cruel sport for those of us in the industry, there is something more important happening here and it's important for the entire marketing and advertising industry.

Back in 1995 when we started in this business, we toured the Internet seminars and everyone was saying the same thing: The world would never be the same again and if you weren't online, you'd soon cease to exist. Apart from a few companies such as Encyclopedia Britannica, this really didn't come true. Some industries changed, but after an initial period where start-ups scared traditional retailers with the size of their initial IPOs, the powerful big companies largely managed to maintain their position despite the entrance of a few Internet start-ups in every category.

However this recent retail issue is quite different. In my opinion it's not about price or GST, although that is part of the equation. Yes you can generally buy offshore for a lot more than 10% off the local Australian price, so even when you factor in freight and the 10% you're still better off. What's it about? Well in my opinion it's about a fundamental change in consumer behaviour.

Internet banking started 12 or so years ago and as we adopted Internet banking we subtly learned a lot of quite complex things. We learned that the Internet could be secure but it could also be insecure. We learned to navigate complex functionality and we learned that it's much easier to do some things online than in a branch.

Most people's first e-commerce experience was with Amazon. Amazon took on a product category many people said was doomed to fail. After all, we loved books and bookshops. However, Amazon offered millions of titles and levels of customer experience that improved on the in-store experience. Reviews, lists, other items bought by people who bought this book, etc. all added to the experience. So Amazon helped people get started with online buying.

Fifteen years later, many people are used to buying online. They know the range is better, they know the experience is better and simple things like "Choose carefully because if you change your mind you won't get a refund" don't exist.

These models have now matured over a long period of time and regardless of the category it's often better to buy online than in a store – so guess what, we're shopping online. My random survey of the shopping behaviour on Sydney's North Shore (conducted, admittedly unscientifically, at the dog park) says most people are buying some, if not all, of their presents online.

We've worked out that we don't need shops when we can buy online and we've worked out that we don't need advertising because we can now find anything we want. For the first time, we're seeing mainstream campaigns that drive no incremental search or web traffic. That's a first. You can spend millions on TV and not see anything in your search or traffic stats. Why? Because everyone who might be interested in your product is already looking for it and they're not hanging around long enough to see the commercial in the ad break.

I think 2010 was the year retailers finally realised the Internet was a harbinger of a change to their business and it'll be 2011 where the same impact is felt across the marketing and advertising business.

The moaning from the retailers is silly. The threat to build stores in China fails to recognize most retailers have missed out on 15 years of learning and there is no way to get that back. It's not about the Internet – it's about a 15-year change in the behaviour of consumers and if you're not ready now, you probably won't catch up.

INTERNET INDUSTRY/ BUSINESS MODELS

NEW MEDIA, NEW APPROACH REQUIRED

(PUBLISHED IN 2002)

Have you ever noticed how important music is for setting a context for a film? Take American Graffiti, for example. If a movie is set in 1962, you expect background music, particularly music played on the radio as the actors drag around town in their cars, to be from the early 1960s.

When it's done right, music appropriate to the era of a film makes you feel you're right there in the middle of the setting. But if they'd played a tune from the 1970s in American Graffiti, George Lucas would have been accused of not doing his homework and the film would have been written off as anachronistic and lacking in integrity.

Deliberate anachronisms – using contemporary music in a historical setting – rarely work on film. Two recent exceptions that I can think of (because I recently saw both of them within a few days of each other) are *Moulin Rouge* and *A Knight's Tale*. The images of Ewan McGregor telling a 19th-century Nicole Kidman that "Love lifts us up where we belong" and a group of 14th century peasants at a jousting match thumping the railing and singing "We will rock you" are jarring at first, but after a while you can see how they help you understand the historical setting by drawing parallels with modern life.

I know, I know, you're thinking, "What on earth does this have to do with Web marketing?" Well, the principle of anachronistic behaviour also applies to different media (although it's not strictly anachronistic behaviour since "chronistic" refers to time, but there's no such term as anamediaistic, so I'll ask you to humour me here). If someone takes the easy way out and simply transfers something from one media to another, you can smell it a mile away. It doesn't work to simply stick a movie or television show on the radio, because you can't see what's happening. Continuing the George Lucas theme, imagine turning on your radio and listening to the final battle scene from *Star Wars*:

Sounds of spaceships whizzing through the air. Explosions. More sounds of spaceships whizzing through the air. More explosions. "Phew, that was close!" *Yet more sounds of spaceships whizzing through the air.* "Use The Force, Luke." *Yet more explosions...* you get the idea.

This is particularly true when it comes to marketing. You don't read out all the text in your DM piece on your television ad, and vice versa. Advertising agencies have built their reputations upon being able to develop the right approach for the right medium.

Then why, oh why do so many ad agencies continue to take old media approaches to their Web work? I see it every day: websites filled with unedited brochure copy and downloadable versions of a company's

TV ads, cool Flash animations that take precious minutes to download while users are waiting to get to the home page and start getting down to business, and sites that show a complete disregard for user-centred design and functionality.

Many ad agencies still don't get the Web. They're still thinking in the traditional advertising paradigm, couching brand strategies in offline terms and relying on gimmicks and giveaways instead of thinking about how a website will change the way a customer interacts with a business.

Using a website as a marketing tool is about usability. From what I can see, many agencies don't rely on usability testing when developing a site. Their idea of usability testing appears to be gathering groups of the target audience in a room and asking them what terms they would use to describe how they feel about a brand. But it's not how people feel about a website that's important; it's how they use it.

Five years ago, if anyone had an opportunity to kick a goal online, it was ad agencies. They had control of customer relationships and they had the imprimatur to develop those relationships in a new medium. But by and large they have dropped the ball. Of the big Web developers worldwide, only a few have their origins in an agency base. Agencies have been good at TV, but the rise of the Web has meant the decline of TV and the decline of agencies.

Why haven't most agencies made the transition? I think it's the fact that agency work is fundamentally centred around the cult of the individual – the power of one creative idea. To do the Web, you need a team involved.

And that team needs to think long term. A website is an ongoing presence; in that sense it's more like a publishing operation than an advertising campaign. The Web is about talking to an affinity group again and again, building a relationship, as happens with a newspaper or magazine. It's about listening to your customers and adjusting your product immediately in response to their feedback.

HotHouse started out as the new media division of a small direct marketing agency. But we realised early on that we were dealing with a very different paradigm in our Web projects. Although we started out building websites and online promotions for the agency's existing clients, pretty soon our client lists started going in opposite directions.

As we gained experience in Web development we developed a set of competencies that were very different from our advertising colleagues. It wasn't just that we were scruffier and we started employing techies who liked to sleep under their desk while completing a deadline. We found we were taking a fundamentally different approach to our work.

The sorts of questions we asked clients during the business development phase were vastly different from what our advertising colleagues were asking. We became familiar with our clients' business processes and not just their marketing functions.

In our case, the Web work became the tail that wagged the dog and HotHouse doesn't do any non-Web advertising work anymore. That wouldn't happen at bigger ad agencies – nor should it – but it highlights the importance of developing a new set of skills to function in this new medium.

Ad agencies still have the opportunity to develop those competencies and blend them together with their current skills to make some beautiful, non-anachronistic music on the Web.

NEW WAYS INFILTRATE THE OLD

(PUBLISHED IN 2001)

Just when it appeared the tide had turned, another wave of tech wrecks has come crashing onto the business shore.

But when we read all those headlines about the collapse of a new industry, we tend to forget that dot coms are only one part of the Internet workforce, which is increasingly made up of clicks and mortar employees. During this year's first quarter – the toughest yet for dot coms – Internet job cuts accounted for only 7% of all layoffs tracked by Challenger, Gray & Christmas. The auto industry laid off nearly twice as many employees. In fact, the University of Texas Center for Research in Electronic Commerce has reported that outside of dot coms, employment in the Internet economy is growing strongly.

Despite both the dot com crash and the current economic downturn in countries like the US and Australia, technology market researcher IDC recently predicted the Internet economy would generate $5.3 trillion in ecommerce sales globally by 2005. "The bursting of the dot com bubble is just the beginning of the market's evolution," reports *The Industry Standard*, revealing that IDC has observed continued spending on Internet-related projects.

What's changing is who those companies are spending that money with. A year ago, specialist consulting services were considered necessary for businesses wanting to be part of the new economy. But many of those consultancies have hit the wall in the last 12 months as businesses around the world have gotten cold feet about turning to companies without a track record run by people in their 20s.

They're returning to big consulting firms like McKinsey, Boston Consulting Group, Accenture, KPMG and PriceWaterhouseCooper for business advice. But it's not quite business as usual for these firms.

The traditional consultants are changing the way they work, operating more like the upstarts they appear to have beaten. Instead of sticking with the traditional pay-by-the-hour engagement, they're exploring alternative revenue models like sharing in increased sales. They're floating their companies on the stock market. They're hiring the top refugees from the failed online consultancies or they're partnering with technology companies (The proposed merger between Hewlett Packard and PwC last year and Novell's recent purchase of Cambridge Technology Partners are two of the biggest examples). Most significantly, more and more of their gigs are designing and implementing online solutions to turn traditional businesses into clicks and mortar operations.

While the focus is still on providing technology solutions, consulting firms are now blending new and old economy solutions.

PROCESS VS. OUTCOME

One effect of this new focus has been to pull IT out of the back room and into mainstream business.

Web-based online solutions are still technology solutions, but they have a much stronger focus on the guy at the end of the chain – the customer. Developing online solutions requires a different way of thinking, a different mindset, than traditional IT-based solutions.

IT consultants tend to be very process-oriented – lots of reports, lots of measurement, no tolerance for deviating from the agreed plan, lots of testing to iron out the bugs before real people can use a system. If you come up with a better way of doing things halfway through an implementation, you'll have to endure the long and painful process of filing a change request and re-scoping and re-costing the entire project.

The Web development process is focused more on outcomes and less on the specific steps to reach that outcome. It starts with a lot of questions: What do you want to achieve? What do your customers want? Do you really understand what you want? It's less "one size fits all", more flexible. This is a reflection of the constantly evolving nature of the Internet. If the goalposts change halfway through, you just adjust the project, you don't force the client to stick with something that was relevant six months ago but may not be relevant today.

For years marketers have been frustrated with the growing influence of the IT department because the technology has always promised so much in the way of customer information and success measurement, but delivered very slowly if at all. The influence of online technology on IT solutions is a godsend for marketers as it forces IT departments to move more quickly and become more flexible.

While marketing has been frustrated with IT because they're slow to move, IT has been frustrated because marketing expects complex solutions overnight and major changes even quicker. Marketers often have trouble defining their needs, other than describing how a system should operate in the end state. They can't identify the most efficient ways of pulling bits together.

It's a classic left-brain, right-brain struggle that has existed within companies ever since computers were invented. The development of ecommerce has added a new dimension to that struggle. In some organizations ecommerce has grown out from the marketing department, while in others it lives in the IT department. Still others have set up ecommerce as a stand-alone division without ties to either department. But in all organizations, marketing and IT are tussling for control over the future of ecommerce.

Who will win? Marketing will eventually come out on top in this struggle, because customers are demanding that businesses move and adapt more and more quickly, that service levels improve, that they anticipate customer needs and don't follow them. They won't accept slow and inflexible processes, even though they'll still demand the quality of experience associated with careful planning and testing.

Especially in tough economic times, if a company doesn't move fast, it's dead.

AN EBUSINESS STATE OF MIND

(PUBLISHED IN 2001)

I was in Chicago earlier this year. A well laid-out city that stretches along the western tip of thumb-shaped Lake Michigan, it has twice the population of Sydney packed into less than half the space. It was summer when I was there, and the CBD was bustling with energy and purpose, thriving despite the slowing US economy.

That's because Chicago, in the centre of the US Midwest's Rust Belt, has emerged as a powerhouse of the post-2000 information economy. Unlike Silicon Valley, which was heavily loaded with dot coms, Chicago is home to a host of old economy companies that are now embracing ways of doing business online, and surveys that include these companies in their figures are showing that Chicago has more people working with new technology than anywhere else in the world.

Walking down Clark Street past the offices of Accenture (formerly Andersen Consulting, the newly public company officially has no headquarters, but as an offshoot of Arthur Andersen has its origins and largest office in Chicago), I could sense something palpable in the air.

Was it the glow of schadenfreude, perhaps, that smug pride in others' misfortunes? Many at Accenture had just gleefully watched a very tall poppy get lopped off at ground level.

Former Andersen CEO George Shaheen jumped ship in 1999 to join startup Webvan, a well-funded online grocery delivery service in San Francisco.

He left at a time when years of unfettered growth by traditional management consulting firms was being seriously eroded by the appearance of "new economy" consultants that were helping take companies online.

Shaheen was the biggest of the big fish who were leaving old economy companies for the brave new world of Internet startups. He was lured by generous stock options and the challenge to build a business from the ground up. Although he left behind a US$4.5 million salary and a retiring partner's pot of gold, he looked like a genius on the day Webvan went public when, briefly, his shares and options were worth US$430 million.

But Webvan's shares never recovered the peak of IPO day, tumbling from US$34 a share even before the tech stock crash. By the time Shaheen exited the company in April this year after 18 months at the helm, it was hovering at just over 10 cents a share, turning him from the $400 million dollar man to the $160,000 man. Any sympathy for Shaheen, however, evaporated when the details of his exit package were revealed: a payout of $US375,000 a year for life. Not bad for a man still in his 50s.

The outcry over his payout and questions about the competency of Webvan management in agreeing to those conditions helped push Webvan's share price even lower and in July it shut up shop and applied for bankruptcy, after churning through more than US$400 million in investor funding.

Former colleagues and competitors chuckled as George Shaheen was forced to join the queue of creditors trying to squeeze some blood out of whatever turnips were left at Webvan.

Meanwhile, Accenture, along with McKinsey, the Boston Consulting Group, KPMG, PriceWaterhouseCooper and the other major management consulting firms, have spent the past year or so re-defining themselves as ebusiness consultants, offering the best of traditional consulting married with world-class online consulting. Trust us for your ebusiness consulting needs, they say, because we'll still be around in five years.

And there's plenty of money to be earned in ebusiness consulting. Despite both the dot com crash and the current economic downturn in countries like the US and Australia, technology market researcher IDC recently predicted the Internet economy would generate US$5.3 trillion in ecommerce sales globally by 2005.

"The bursting of the dot com bubble is just the beginning of the market's evolution," *The Industry Standard* (a publication that itself became a victim of the crash) recently reported, revealing that IDC has observed continued spending on Internet-related projects.

There are just a couple of problems with the move of big management consultants into ebusiness. One, many specialist ebusiness consultancies are still in business – while Internet pure-play businesses have gone to the wall, traditional companies are lining up to expand their business online. And two, those specialists have turned traditional consulting models on their head, and clients are happy with doing business in these new ways.

Ebusiness developers, because they had their origin as small, passionate users of technology to solve a big problem, have been willing to stick their necks out and offer fixed fees that depend on whether or not the project is a success.

Management consultants, meanwhile, most of whom have their origins in large accounting firms, enter open-ended, no-risk arrangements. It's like getting in a taxi and leaving the meter running, except you don't know where you're going and how long it will take to get there. The taxi driver is going to go the route he chooses to go and keep the meter running until you decide you want to get out, regardless of whether you're happy with the destination.

That approach just doesn't work in ebusiness – at least it doesn't work for the client.

In ebusiness consulting, you need to enter an agreement of trust with your client – trust that you won't finish until everyone's happy. You can't just do a job as quoted and walk away, pocketing the money for time spent. You have to come up with a solution, but be willing to embrace uncertainty and the possibility that circumstances may change from the beginning to the end of the project, and not leave until it's working to the client's satisfaction. And satisfaction doesn't mean throwing money at extra consultants and handing the client the bill.

Ebusiness consulting is not a case of installing the latest CRM, CMS, ERP or ERG system, shaking hands and leaving. It means crafting a practical, flexible framework, and sticking with a project until the client's outcomes are achieved. Ebusiness consultants position themselves as partners, and do whatever it takes to meet their clients' needs.

Web-based online solutions are still technology solutions, but they have a much stronger focus on the person at the end of the chain – the customer. Developing online solutions requires a different way of thinking, a different mindset, than traditional IT-based solutions as practised by management consultants.

Management consultants tend to be very process-oriented – lots of reports, lots of measurement, no tolerance for deviating from the agreed plan, lots of testing to iron out the bugs before real people can use a system. If you come up with a better way of doing things halfway through an implementation, you'll have to endure the long and painful process of filing a change request and re-scoping and re-costing the entire project.

The Web development process is focused more on outcomes and less on the specific steps to reach that outcome. It starts with a lot of questions: What do you want to achieve? What do your customers want? Do you really understand what you want?

It reflects the constantly evolving nature of the Internet. If the goalposts change halfway through, you just adjust the project, you don't force the client to stick with something that was relevant six months ago but may not be relevant today.

John Ellis, writing in *Fast Company*, sums it up well. "The traditional consulting model no longer applies. It isn't even close to the new reality of the Web. The traditional model said, 'We make a trade: your money in exchange for our smarts.' The Web model says, 'We collaborate – and get smarter together as we go along."

The traditional consultants, from their headquarters in Chicago, New York and London to their outposts all over the world, are walking tall in the post-crash business world. But it's not a time to be complacent. If they don't adapt to Web-based ways of thinking and working, they'l fluff their ebusiness consulting opportunity.

INTERACTIVE AGENCIES NEED TO STOP BEING ADVERTISING AGENCIES

(PUBLISHED IN 2009)

I had a bad new media week last week, a bad interactive media week, and a bad social media week. It's been coming for some time but it all culminated with ad:tech.

I've been in the interactive media industry since 1994 when we were making CD-ROM titles for the US market. Like many others, I was attracted to the industry because we were given an opportunity to be at the forefront of a new technology. We were in a position to challenge the status quo and to do something that mattered.

As an industry, the last 15 years have been extraordinary. The world has changed. Minnows attacked huge existing businesses and won. In some cases, those traditional businesses are still around. But in many cases, they are not. There are many examples:

- Why did Brittanica not become Wikipedia? They had a start, they had the content.
- Why did LexisNexis not become Google? They had a start, they had the content.
- Why did Trading Post not become eBay?
- Why did Dymocks not become Amazon?

Too lazy, too stupid, not ready to take a risk?

And the examples go on. Small businesses have in a short time overtaken and beaten traditional business by changing the game. To be part of all that has been a privilege.

So, what started this rant for me? Some of it was the plethora of new jargon at ad:tech – some of it was lameness of the examples, and some of it was reading *B&T* again. I think the interactive industry wants to be the advertising industry, we think it's cool. They have great offices and the parties are fun but they work people like dogs, get treated like shit by their clients and we want to be like them.

Here are a few things that got up my nose:

Brand dialogue: What a load of crap. I've worked out what it means. It's a Flash site with some un-navigable metaphor and takes ages to download. The interactive industry despise them because we know they are a waste of time. Most of them get a campaign URL and Google does not take it seriously because it's new.

Advertising agencies tell their clients it's part of the brand dialogue and measure hits to the websites. We never hear bounce rates quoted; that would be embarrassing. Agencies tell clients they can do some things to help get the Flash site indexed by Google. But we're not really telling the truth. Agencies build them because they are the closest thing to a commercial. They understand those things and clients seem to like them because they're familiar.

Companies succeeded on the web because they solved a problem: they found a way to facilitate a transaction. Even Apple, the high-water mark of the "brand dialogue" debate, just facilitates transactions. Everything they do makes it easy to transact. Despite all the hoo-hah you read about corporate social media, consumers don't want to talk with brands – we just want the Internet to facilitate an easier way to do business.

We, the interactive industry, know this, we can see it in the numbers. Why don't we adopt our own language and help our clients build business on the web?

Social media: I was at a dinner party. The people there all used eBay, bought music online, had laptops, iPhones and used BitTorrent for downloading TV series. I asked if they knew what social media was. They did not. Sites I called blogs they called websites. As far as they are concerned Facebook is a website, LinkedIn is a website, MySpace is a website.

PR agencies are distributing press releases to bloggers and calling it social media. Banks are getting 50 comments on a blog and talking about it. In the old days, you took your clients to a lunch and got a better return.

In 1997 we built a website called Manhood in conjunction with psychologist/author Steve Biddulph. We built in every feature you see in a social media site. Then, it was common sense; now, it's social media.

As marketing people our job is to go where the customers are and find a way to tell them about our product. We in the interactive industry know where these people are to be found, we know how to engage with them and we do really know what social media is, but we're dumbing down our business is an attempt to look like an advertising agency.

Posting a few images to Flickr, the TVC to YouTube and putting the URL on the TVC is not really going to work.

Then there was the Naked stunt for Witchery. The only interesting thing about this was the fact the *SMH* chose to publish it on the home page. The interactive industry knows the only measure of success here is, "Did you sell any jackets?" That's been the mantra of our industry – what's the outcome – and yet we're swept up into a cycle of debate on the topic. It's presented as an Internet/social media strategy, but it's neither.

Let's go back to what we know is true – if it works by delivering commercial value to the client it's a good thing. Let's leave advertising agencies and PR companies to the rest.

FMCG: Every second person we interview wants to talk about our FMCG clients. Well, we don't have many. Now Skittles is kinda interesting, but really who cares. It's a sweet. The purchase decision is made in a store and this is more interesting to the industry than the target market.

There was more traction from the guys putting Mentos in Coke than any of this stuff and we, the interactive industry know this but we want to pretend to be advertising agencies. When I was a kid you had cereal box competitions - collect tokens and win etc. Most FMCG websites are nothing more. They are not cool, they are not clever they are not worthy of debate, but they get written up and talked about at conferences like they matter.

Metrics: A friend of mine stopped reporting clicks, conversions, etc. to her client because it raised more questions than she could answer. Clients have been pumping money into TV for years without any real sense of a return. All they ever asked was how did the show rate.

Sometimes we can't quite measure the impact on sales. There is often a lot of air between the website and the sale, but we can measure everything else. We, the interactive industry, know how to do this. We've worked through the issues for years, but we're allowing ourselves to be dragged down the reporting on ratings path when we know we can do better.

Media planning: Rex Briggs, the king of measurement, says most organizations should be spending 20% of their budgets online. It's not happening. Media buyers run the same ads on the major portals despite the fact the interactive industry knows creative wears out quickly. And why don't we make more creative? Because the media buyers do a deal with the media companies for 16 sizes to fill all sorts of stupid holes. Instead of running two sizes and rotating creative we get 12 sizes and blow our budget. The interactive industry knows this is wrong.

Why are we on these portals, anyway? Sometimes it makes sense, but most of the time it's because it's too difficult to do a proper job. The real engagement is down the other end of the traffic curve - the niche sites where people are spending time and returning regularly.

Media people will tell you it's because they want their brands associated with the prestigious brands. What a crock – the interactive industry knows it's a crock. Have you seen the "quality advertisers" filling the unsold inventory on the major portals? It makes 3am TV look like quality.

Advertising Agencies: Every single agency in town had the opportunity to own the interactive space. They had the clients they had the mandate and they blew it. The top 10 digital services companies in Australia would, I suspect, all be web companies, not agencies. Just like the Trading Posts of the world, the ad agencies did not grab the opportunity to change.

Some larger organizations use their procurement departments to run the tender and selection process for the advertising agency. This puts agencies in the same league as stationery, toilet paper and the company car fleet. They would not dream of doing the same for the legal partner, the strategic consulting partner or the HR consultants.

The interactive industry can avoid this horrible fate, but we need our own sense of pride and culture. We don't want to talk like agencies, we don't want to aspire to be part of their magazines; why would any digital person want to be on the Gruen Transfer for example?

It's time for our industry to re-establish an identity separate from the agencies, to make sure the work we do delivers a real return and to work on things that use our energy and intellect to make a difference for our clients.

I'd like to propose an Interactive Industry Code of Practice:

1. I will always propose the least expensive, simplest solution to any problem.
2. I understand Google is the homepage and I will ensure everything I do is sensitive to this fact.
3. My job is to facilitate business. When I start talking brand dialogue it's only because I can't find a way to really add value.
4. My job is to help you with the interface between your company and the customer on the web. They are using the web for utility; my job is to find that utility wherever it may exist.
5. We'll be clear about the returns.

6. We have a chance to do things better to improve from our learnings.

7. The Internet has changed the world; let's make sure we treat it with the respect it deserves. It took us many years of TV to develop the technology to skip ads. Let's not clutter our communities and forums with useless messages that add no value. Consumers want to hear from companies who are relevant to their circumstance; let's work with that.

THE MORE MARKETING CHANGES, THE MORE IT STAYS THE SAME

(PUBLISHED IN 2010)

The term "legend" doesn't apply to many people in marketing in Australia, but one man who clearly qualifies is Bob Miller.

Long-time general manager of marketing for Toyota Australia, Bob is best known as the man behind the classic "Oh, What a feeling" campaign that placed Toyota at the top of the Australian car market.

Bob is one of Australia's leading authorities on sales and marketing and numbers among his many awards the Australian Marketing Institute's Sir Charles McGrath Award, Marketing Magazine's Marketing Executive of the Decade and BRW's Marketing Director of the Year.

Now working as a speaker and consultant on marketing and Internet business development, Bob has written a number of books and magazine columns on marketing and is also an adjunct professor of business at Macquarie University.

I had the privilege of interviewing Bob for a recent podcast. When I asked him about the biggest change to advertising and marketing in Australia during the past 30 years, he didn't nominate the rise of digital marketing, but the end of the fixed margins and commissions of the media accreditation process in the 1990s.

During his tenure as president of the Australian Association of National Advertisers, Bob Miller was instrumental in the Australian Competition and Consumer Commission calling an end to the media accreditation system. Started by News Corp's Rupert Murdoch in the 1960s, the system allowed media owners to select the firms which could receive payments (some called them kickbacks) of up to 12% for placing advertising, while advertisers were excluded from such deals.

He made a few enemies in the process of pulling down the accreditation system. Several years down the track, he rates the outcome as "disappointing, because agencies haven't taken advantage of the situation, while media houses have."

MARKETING AS SYMPHONY

The job of marketers, Bob Miller says, is similar to conducting an orchestra. "It's the conductor's job to take all the individual musicians' activities and turn them into Beethoven's 5th Symphony. That's the marketer's job, though whether they get on with it is another thing."

Asked whether the fundamentals of marketing have changed in the past two decades as a result of the digital revolution, Bob is emphatic. "No! Marketing hasn't changed since the markets of ancient Assyria. 'Buy my oranges, they will last longer, taste better, etc.' The same applies today.

"Marketing may be more complicated, but it essentially has not changed. Markets are conversations. The ideas are the same."

Social media, he says, like everything in its early phase, "is over-hyped – but it is meaningful. It's evolving."

At the end of the day, according to Bob Miller, whether you're using digital or traditional marketing, "it's about generating predictable future cashflow for the business and shareholders. The job of marketers is to convince the board to give you money so you can generate that cashflow.

Digital technology, he says, "complicates the life of marketing directors," particularly because most CEOs and CMOs are still firmly entrenched in TV, newspapers and direct mail, three media they grew up with and understand.

His advice to marketing agencies of the future is to address the cashflow issue by embracing statistics. "You need to have numeric insight, to sell your ideas to shareholders and stock market analysts.

"Agencies who service business accounts need to have conversations with econometricians – or hire some young ones with this background. You need to demonstrate to a company board that here's a new way of thinking about the future, the future of your cash flow. That's why we do advertising."

For companies to be successful in the future, Bob reckons, "The finance department should be integrated into marketing department. I know the CFOs think it should be the other way around, but they need to work more closely together understand that it's a marriage of interests – and the stakes are high."

MARKETING AS A CONVERSATION

THE PAST OFFERS THE CLUE TO THE FUTURE

(PUBLISHED IN 2001)

Despite our Internet take-up rate being among the highest in the world, Australians have been notoriously slow to embrace ecommerce – maybe it's because we're too busy looking at pornography online.

In the lead-up to Christmas 2000 less than 15% of Australians visited a shopping site online, compared to 50% of Americans (AC Nielsen figures). Canada, with comparable Internet use levels, had 25% of its population transacting online during 2000, while Australia reached only 12% (Jupiter Research figures).

But don't worry; there is an online category where Australia puts in a gold medal performance – viewing smut. A Media Metrix survey has estimated one-third of Australia's 6.8 million home users checked out pornography sites during December, placing us equal first with Canada and ahead of those repressed Americans and Europeans.

We can only hope those millions of Australians were multi-tasking, which you can do at sites like Nakednews.com, where the news, sports and entertainment streams out of your PC's speaker while the newscasters' clothes hits the floor.

ECOMMERCE MORE THAN JUST $$

I think Internet analysts are placing too much emphasis on dollar transactions online as a measure of development of the market. Some of the most important ecommerce activities aren't even happening on corporate websites, and they are having a huge impact upon business as we know it.

These behind the scenes changes are chronicled in a website and book called *The Cluetrain Manifesto* (or more accurately, *the cluetrain manifesto*, displaying the new economy penchant for lower case letters), written by four Internet pioneers who have taken conventional business theory and thrown it in the bin.

The manifesto is comprised of 95 theses – reminiscent of Martin Luther (the 15th century German theologian, not the black American civil rights activist) and his ideas that challenged the Roman Catholic Church and brought on the Reformation. The Internet, these fellows argue, will bring about changes just as big.

It's never explained exactly what is meant by the word cluetrain, but I think it means something like "the Internet train is leaving for the future, and if you've got a clue you'll hop on board".

The main premise of the manifesto is that markets are conversations. Long before big business rose up and started driving the economies of the world, people used to conduct business through conversation, discussing the weather, family and politics before finally bartering their way to a deal.

What does this have to do with the Internet? The manifesters are arguing that the most important facet of this most modern of technologies is that it enables people to have conversations directly with each other. The effect for business is a return to the traditional marketplace. As they write, "In many ways, the Internet more resembles an ancient bazaar than it fits business models companies try to impose upon it."

SUBVERTING BUSINESS HIERARCHY

By using online technology – intranet as well as Internet – people in networked markets are finding that they get far better information and support from one another than from "official" business. Or, as thesis 7 puts it, "Hyperlinks subvert hierarchy".

Ebusiness pundits – myself included – writing about the online revolution have placed the focus almost solely on the public side, B2C and, more recently B2B and B2I (business to intermediaries). But we also need to consider the fundamental changes the Internet is bringing about inside businesses.

The questions businesses should be asking themselves have to be not just "What does my website need to look like?" but also "How can we use online technology across our entire business?"

This is not life at the sexy "bleeding edge" of the Internet (and hasn't there been a lot of blood spilt at the edge in the past year?), but it is still a revolution nonetheless.

There's a lot happening online that's affecting your business, even if you don't have a website. It's happening in places like your intranet, in email conversations between your employees and between your employees, customers and suppliers.

As the cluetrain manifesto points out, staff at all levels are having contact with each other and with customers where they never did before. It's informal and it's not easily controlled. In fact, attempts to control it will backfire miserably. Businesses need to loosen up and let these conversations happen.

I don't agree with everything the manifesters put forth. Their assertion that employees surf bulletin boards and email lists on their own time looking for mentions of their company's products and responding to problems in questions on behalf of their company is hard to believe – at least in Australia.

But the mere existence of these lists and forums, where people step in to share knowledge and gossip, shows that there are thousands of conversations going on that could never have happened without the Internet. It is genuinely empowering people, and even if businesses feel threatened by those conversations, it can only benefit the business community in the long run if they listen to what their customers are saying and respond.

The pendulum swing toward listening to customers that began with the total quality management (TQM) movement of the 1970s is swinging stronger than ever because of the Internet. But what a scary concept it is that listening to customers is a revolutionary business idea. Isn't that what we're in business for?

CONVERSATIONS ABOUT CONVERSATIONS

(PUBLISHED IN 2008)

On the Internet, anyone can become a media mogul. Unlike traditional media, the set-up costs are practically zero. Of course, gaining an audience is another matter entirely. You need to have something worthwhile to say, say it in an interesting and articulate way and know how to gain attention.

As a result, hardly any new media publishers are making serious money. The exceptions, such as The Huffington Post in the US and Crikey in Australia, have been started by refugees from traditional media. The splintering of audiences caused by the Internet makes it difficult to consolidate eyeballs, and therefore revenue.

However, this media fragmentation has led to the rise of thousands of niche publishers who bring their own personal brand of information to their specific audience. When it comes to chronicling the development of the Internet, one of the most widely-connected independent publishers is Susan Bratton, whose Personal Life Media 'empire' features the weekly podcast "DishyMix: Juicy Interviews with Famous Internet and Media People".

Susan was a founding member and vice-chairman of the Internet Advertising Bureau and launched online advertising products for companies such as AOL and Excite before starting Personal Life Media. As a result, she has met many of the big thinkers in online marketing and advertising, and she has combined her address book with her engaging interview style to produce content-rich interviews with scores of Internet entrepreneurs, CEOs and executives. DishyMix provides a great helicopter view of new media, where it's come from, and importantly, where it's heading.

Though the interviews cover a broad spectrum of new media, most coalesce around four interconnecting themes. Here's a summary of the latest thinking around those themes, as espoused by DishyMix guests:

1. SEARCH

- Google has become the world's home page. As a result, search is now a global activity, which creates issues for local and national brands. Channel conflict abounds as head offices compete with branches and manufacturers with their distributors in the search arena.
- John Batelle, a blogger and digital entrepreneur who wrote the definitive best-selling book about Google, *The Search*, says that Google proudly declares that it is not a media company, but it needs to start thinking like a media company because of the place it now occupies.

One of those things, like it or not, is the cyclical nature of the media business. After not having had a bad day in 10 years, Google is caught up in the current financial downtown because, if people are buying less stuff, they're searching for less stuff (following this logic, look for eBay to have a resurgence as people sell all that stuff they can't afford any more).

- Search engine marketing specialist Danny Sullivan, meanwhile, predicts that search marketers will need to change their strategies to accommodate 'blended results' as maps, video and local information are added to search results.
- Search engine optimisation expert Stephan Spencer, who runs Netconcepts, says that link baiting – publishing a definitive story or blog post on a topic that everyone wants to point to – is the simplest way to drive traffic to your brand.
- DishyMix guests agree that SEO people make the best social marketers, because, as Susan describes, they have a sense of "who am I talking to, and how do I elicit a response? What's a tag that people will jump on?"

2. SOCIAL MARKETING

- Social marketing is more than just posting your latest commercial on YouTube. It is a fundamental move in marketing strategy from interruption and push to invitation and engagement. It is turning a prospect on to a brand idea enhanced by the surrounding context.
- Digital media is not about buying ad space, it is focused on distributing experiences. The key is to create something that pulls people together and gives them something to do.
- Charlene Li from Forrester Research, who helped articulate the Social Media Participation Pyramid, reports that the pyramid is

changing shape as more people move from being spectators (at the bottom of the pyramid) to creators (at the top).

- Trying to have a viral marketing show is almost impossible. Viral is not a strategy; it's a lottery.
- Despite the bad press it received in the wake of the tech stock crash, the first mover advantage is alive and well in social marketing. The General Motors blog, written by the firm's CEO, enjoyed outstanding press coverage, or as Susan calls it,"tremendous value in earned media".
- Peter Shankman, PR entrepreneur and creator of the wildly successful Help a Reporter Out project, says you need to think about who you are and what personas you need to hold online. If you're a prominent executive, do you publish photos of your holiday on a dude ranch in Montana where people can find them on your Facebook page? And how do you define 'friends' on sites such as Facebook and MySpace?
- Stephan Spencer says you should leverage your presence on LinkedIn by using tactics such as including your email address in your name listing, connecting to a LION (LinkedIn Open Networker), and making sure your entry includes links to your blogs, web pages, etc. to increase your own Google rankings.

3. MARKETING AS CONVERSATION – THE ART OF BLOGGING

- The term has been bandied about ever since the publication of the paradigm-shifting book *The Cluetrain Manifesto*, but what does 'marketing as a conversation' really mean? Susan describes this turnaround in marketing activity and power as "going from scheduling media and blasting out to everyone, to consumers, through consumer-generated content, being equal to what we can do as marketers."

Instead of companies with customer databases, customers now have databases of the companies they want to deal with.

- The best way for a company to start a conversation with its market is by blogging. Rohit Barghava from Ogilvy PR, author of *Personality Not Included* says that a corporate website doesn't have much personality, but a corporate blog helps you create personality around your brand. A blog goes beyond authenticity to the personality of your corporation.

- Susan Bratton says a common refrain from her interviewees is that "A corporate blog is the first thing that most companies can and should do to put their toe in the water" of social marketing. The fundamentals of corporate blogging are:

- You need to have one central point which consolidates all your company's blog activities and shows your tone of voice.

- Make sure more than one person is involved in your corporate blog.

- Establish an editorial calendar, a certain time each week where you commit to updating the blog which is, as Susan puts it, "baked into people's calendars'- that way you'll avoid 'blog fade'"

- Lots of customers use your products in unique ways – get them to blog for you on your site.

- The other side of blogging is monitoring outside blogs in your area of interest. You need to identify bloggers who are popular, respected and writing about your brand because, as e-marketing expert Seth Godin says, "Bloggers are the new gatekeepers". You need to become familiar with the tools that can help you find this information, such as Google Alerts, Yahoo!'s Site Explorer, Technorati and Trackur.

4. NUMBERS/ROI

- The great promise of the digital media is that everything can be measured. But what should you be measuring, and what is easy/cost-effective to measure? Rex Briggs, who runs the Marketing Evolution website, argues that ROMO – return on marketing objective – is a more effective measure than ROI in the digital space.

ROMO compares a common measure of advertising or sales response across a range of media. Crossmedia surveys and research reports measure the value of the synergistic approach to media. It allows companies to measure 'uptick' in total value of a campaign mix that models in interactive.

It also includes measurement for intangibles such as engagement, mindshare, not just purchase intent and traditional measures.

- Charlene Li from Forrester relates that her book *Groundswell: Winning in a World Transformed by Social Technologies* contains a formula that allows you to estimate the value of doing a corporate blog, so you can win over internal stakeholders.

- Sean Cheyney from online insurance broker AccuQuote.com uses the measurement potential of the Internet to take A/B testing to a whole new level, with his use of multivariate testing on landing pages. He claims to use as many as 230 iterations of a landing page – some with only nuanced changes – to determine which ones convert the most good leads.

According to Susan, nearly all of her interview subjects have two things in common: 1) They have no regrets – they don't look back, they don't focus on what they've done, but on where they're going; and 2) They all believe in taking risks, that playing it safe doesn't pay off. Considering the calibre and achievements of the people interviewed on DishyMix, I'd say that's a good recipe for digital success.

MEASUREMENT/ DATA ANALYTICS

A THREE-HOUR TOUR OF WEB MEASUREMENT

(PUBLISHED CIRCA 1998)

As if it wasn't hard enough remembering all the normal advertising measurement jargon, along comes the World Wide Web with a whole new vocabulary and an entirely different way of counting bums on computer chairs. While trying to get a handle on online advertising tools and terminology, we came across a transcript from an online Webvertising course conducted by Internet chat. That transcript is reproduced here. (NB: User names have been changed to protect the innocent.)

> Professor: Sit right back and welcome to another session of Webvertising 101. Today's topic is Web measurement. Who can tell me the most important way of measuring traffic to a Web site?
> Gilligan: Hits, sir?

Professor: Where have you been, Gilligan – on a desert island? Hits as a useful measure are ancient history. It may sound impressive to say your site is being hit half a million times a month, but keen students of Web measurement know that hits refers to the total number of text and graphic files on each page. For example, if you have a block of text, five photos, 12 graphic headers and 32 buttons on your home page (as many Web sites do), that counts as 50 hits every time that page is downloaded.

Some unscrupulous Web site developers load their pages up with buttons and graphic headers to boost their hit rate and artificially inflate their advertising prospects. As a result, hits are a totally unreliable way of measuring how many people are visiting your site or what they're looking at while they're there.

Mary Ann: Then how do you work out whether your Web page has made an impression on Web surfers?

Professor: Very good, Mary Ann: Impressions is exactly the right answer.

Mary Ann: It wasn't an answer, it was a question...

Professor: Whatever. Impressions, or page views, are the number of times a page has been served up to someone. They are the closest equivalent to eyes on a page. Compared to the guesswork involved in calculating how many people looked at a particular page in a newspaper or magazine, or how many people were in the room paying attention when a radio or TV ad was broadcast, impressions can be tracked precisely using the Web measurement software packages available to any Web developers.

Ginger: You described impressions as "the number of times a page has been served up to someone." How do you know who that someone is?

Professor: Web measurement software can tell you where the people visiting your Web site came from – sort of. Each time someone logs onto the Internet, their service provider gives them a temporary Internet Protocol (IP) address which follows them around their online wanderings. Web measurement software captures these IP addresses and reports each unique one in its analysis.

The problem is, if someone's using their company's network to access the Internet (which is happening with more companies all the time), everyone on that network will have the same IP address. If 20 people from one network visit your site one day, the data will show that one person visited your site 20 times.

The other shortcoming with Web measurement tools at the moment is that they don't give you any demographic data on the people visiting your site. The only way to get that is to convince people to register with your site and, in the process, answer a few questions about themselves.

Skipper: I thought being anonymous was one of the most appealing features of the Internet. Why would anyone want to tell you details about themselves if they didn't have to?

Professor: That's one of the biggest dilemmas facing Web developers and advertisers at the moment. The trick is to give people something in return for that information, like a discount or a subscription to a useful online newsletter. Some car makers, for example, are offering $100-$500 off the list price of a new car to people who register with their site and fill out a detailed survey.

Mrs. Howell: How are Web publishers using this information to work out how much to charge for an online ad?

Professor: Publishers are using a combination of different methods in determining online advertising rates. In Australia, most online advertisers are being charged flat monthly rates, with the highest rates going for banners on home pages. The high-traffic international Web sites are using impressions as a basis for a CPM-type rate charge – the more times an ad banner is loaded onto a user's page, the more the client pays.

Thurston Howell III: Measuring eyes on an electronic page is all fine, but isn't this medium supposed to be interactive? Can't you measure whether people do more than simply look at an ad on a page?

Professor: Yes, Web measurement tools not only record how many times a page is served up, but also which bits, such as the ads, are clicked upon. In the case of advertising banners, you can determine how many people clicked on the banner – called click-through – and you can track their journey through the advertiser's Web site.

Thurston Howell III: I like the sound of that. If I can work out how many people actually travel outside the site to read my ad, then I can screw ad rates to the wall by paying only for the number of times my ad was clicked on. Paying on impressions is like your salesperson telling you she made 200 sales calls that day. Who cares? How many sales did she close?

Professor: There are a few problems with that approach, Thurston. Among other things, it puts an unfair onus on the Web site publisher. What would you say to a publisher who achieved poor click-through because your ad had no call to action or was boring, poorly designed, the wrong colour, placed in the wrong kind of site for the demographic, or linked to a corporate site so stultifying that no one would ever consider coming back for a second visit?

Thurston Howell III: Tough luck?

Professor: I don't think so. Web publishers need to work closely with advertisers to make sure the Web works for them both, but they can't take on the lion's share of the risk, including many factors which are beyond their control.

Mary Ann: So, at the end of the day, how useful is Web traffic measurement?

Professor: The information available to Web publishers and advertisers is far more precise and scientific than the measurements used in any other medium. If a Web developer or an advertiser wants to know how many different users surfed through a particular Web site, which pages they saw and even how many times they clicked on an advertising banner, it's all at their fingertips.

However, because the Web is still in its early stages of development and things like ad sizes and measurement terms are nowhere near standardised, it's hard to know what combination of information provides a useful indication of how many and what sorts of people are visiting a Web site.

The trick is not just knowing where people came from, but why they came to you. What were they looking for, what was the need they wanted to fulfill? When evaluating the data coming from Web measurement tools, you need to decide which bits of information are useful and which bits should be cast away – every case is unique.

Speaking of castaways, it seems like some of you have been drifting off and haven't asked a question for a while. Gilligan, what about you? Gilligan? Gilligan! Wake up!

(Sound of a virtual duster being thrown across a modem line and landing on the side of Gilligan's head)

Gilligan: Ow! That hit certainly made an impression on my scalp and it's caused a bit of exposure, too!

Skipper: Are you OK, little buddy?

TAKE WEB MEASUREMENT STATS - PLEASE

(PUBLISHED CIRCA 2000)

My New Year's resolution this year was to not take things quite so seriously, to laugh more. It's only appropriate, then, to write about website measurement and online usage statistics, because they are beyond a joke. Several years into the Internet revolution, it's no easier than it was in the early days to come up with reliable figures.

If they can invent online technology that recognises me every time I return to a site (by offering me something enticing, like a cookie), why can't anyone tell me accurately how many people are online, or how many visit a particular website?

Note that I didn't say tell me how many people are online; I said tell me accurately how many are online.

Estimates of the number of Australians online vary from less than 2 million to more than 4.2 million, while the US audience is estimated at anywhere from 37 million to 65 million, and is predicted to climb to anywhere from 90 million to 120 million by 2002. Online shopping is due to climb from US$5 billion to as much as US$37.5 billion during the same period.

But with that kind of variation in estimates, how accurate are the figures? A big part of the problem comes down to how you define the term 'Internet user'. Depending on whether you're talking to Jupiter, Forrester, Nielsen or the Australian Bureau of Statistics, it ranges from anyone out of nappies who has walked past an online screen during the last 12 months to people over the age of 18 who have used the Internet at least three times in the last three months. I tend to believe the latter definition is the most useful, but the former makes for bigger headlines: 'More people online than on the phone!'

One reason why we're having so much trouble coming up with a standard definition of an Internet user is that everyone is looking at the Internet with their TV hat on. The fact that we use a term like 'audience' shows that we're trying to use an apple to describe an orange.

It's only natural that we're trying to define the Internet in terms of what we're already familiar with. But while it may be natural, it's wrong. The Internet is unlike any other medium that has ever developed. The only thing it has in common with television is that both are viewed on a screen. TV is the classic 'cool' medium, as Marshall McLuhan wrote more than 30 years ago, a low-participation, passive experience (except for rigorous channel surfers). The Internet, on the other hand, is a 'hot', high-participation medium, becoming hotter every day with the new multimedia developments.

I think one of the first things we need to do is get rid of the word 'audience' when referring to people who use the Internet, and replace it with a more descriptive term, such as 'participant'. The number of participants on the Net may not sound as impressive as the audience numbers being touted at the moment, but they'll be much more useful for marketers.

EASY AS FALLING OFF A LOG FILE

There's a new joke travelling around online circles (at least it will be, as soon as I start circulating it): How many Web traffic measurement code-cutters does it take to change a light bulb? No one knows, because every time they do it, they come up with a different number.

What, you're not laughing? Neither is anybody who is trying to work out the effectiveness of their online message. The lack of anything resembling a standard in website measurement means that marketers at both ends of the spectrum - those looking to buy advertising space and those trying to find a way of making sense of the log files they've compiled on their website - are finding it nearly impossible to determine if they're getting enough bang for their buck.

The old hits vs. clicks vs. page views debate is still raging. At the same time, website traffic measurement programs have become unworkable and unintelligible because users bounce around different areas of a site and search engine spiders and programmers performing maintenance on a site are counted as potential customers. As Geoffrey Moore, writing in Crossing the Chasm, says, "This stuff is like sausage - your appetite for it lessens considerably once you know how it is made."

Time will tell whether a workable standard emerges from the quagmire of measurement software, or whether website owners will come to rely on other ways of measuring the effectiveness of their website, including old-fashioned methods such as adding up how many more people buy their product or service.

ONLINE ADS - IS ANYONE HOME?

Like the adage about a tree falling in the forest, online marketers are now asking: If an ad banner is splashed across a bunch of websites and no one clicks, did it make an impact?

Potential Web advertisers are keen to know the answer to this question, because the latest figures from the US survey company NetRatings show that the click-through rate on banner ads is only 0.5% - once every 200 page views - down from 1.3% six months earlier (you can probably believe these figures because they admit bad news).

That's not to say that banner ads don't work; it just puts paid to the argument that click-through rates (or CTRs, as they're coming to be known) are a useful way of measuring the success of an online ad. As with traditional advertising, there is a value in simply having your advertising message seen, even if the target doesn't do anything about it right away.

Figures from the Internet Advertising Bureau and Millward Brown Interactive back this up, showing that ad awareness without click-through is nearly as high as ad awareness with click-through.

AUSTRALIAN SILENCE

Of course, what makes quantifying users, website traffic and banner traffic even tougher for Australian marketers is the paucity of local online data. There just isn't enough critical mass for Australian online research companies to justify releasing Web traffic and ecommerce statistics for their publicity value.

This begs the question: How can you decide whether it's worth building or extending a website or buying online ad space? In the absence of hard data, how can you do a cost-benefit analysis?

The answer is, you can't, at least not today. Cost-benefit analyses are biased toward existing media, which operate in a straightforward, one-way message framework. You send a brand message or a call to action to potential customers. They may send a message back by buying your product or service or feeling more favourably disposed toward your business.

But on the Web, marketing messages extend into new possibilities presented by technology. Walid Mougyar, writing in Business 2.0, calls this "turning the unthinkable into the irresistible". The ultimate goal of your website or your online marketing message is much richer than simply sending out a message. It's about providing services that are simply not available through the physical part of your business.

Simply put, you need to get your head around how you can use online technology to enable your customers to do something they can't do anywhere else. As Mougyar writes, "Make your customers comfortable first, then startle, surprise and delight them. Ultimately, companies will be judged not by what their customers can do on the Web, but by what they can do only on the Web."

The businesses that put aside the spreadsheets and the log files and focus on giving their customers an online experience that surprises them and develops their relationship with them are the ones that are making a success out of the Web, regardless of how many people are out there or how many are visiting their site. And that's no joke.

THE UNBEARABLE LIGHTNESS OF MEASUREMENT

(PUBLISHED IN 2002)

> *"We have a global network, broadband, and data centers. Now all we need is a way to make money."*
> **– Technology Brief no. 3, from "20 Technology Briefs", Fast Company**

One of the cleverer circulating emails that pop up in my inbox every few months is the game card for "buzzword bingo". Take it into a meeting and you can tick off words on your card as people in your office use phrases like "B2B", "let's all sing from the same song sheet" or "intellectual capital".

A new phrase that's been getting a bit of a workout on the buzzword bingo card lately is ROI – not "return on investment", but "return on Internet". In the post-dotcom landscape, businesses want to be able to measure the payback they're getting from the money they spend on online projects, and to get that payback quickly.

This is simply the latest of many attempts to put the Internet in a box it was never designed to fit into. Instead of stepping back and letting big-picture growth happen, they're trying to squeeze out short-term returns.

> *"Technology doesn't make you less stupid; it makes you stupid faster."*
> **– Technology Brief no. 7**

The Internet can't be expected to behave like a traditional business process. It's a transforming technology. There is no clear-cut way of determining how much money a business will earn or save when it brings an offline service online, as you can with a new assembly plant or a new warehouse.

Mind you, the Internet isn't alone in being hard to put your arms around. The same could be said for just about any technology solution of the last 30 or 40 years. Ask your financial controller whether he or she can demonstrate how much money your CRM or your SAP system has saved your company.

For that matter, ask your financial controller to justify having a PC on everyone's desk. Sure, everyone knows using computers saves you money, but try to show it, taking into account product obsolescence, upgrading to the latest version of Windows, Microsoft Office or Internet Explorer, replacing bung keyboards and mice, and maintaining an IT department. The promise of technology-driven productivity gains is a self-fulfilling prophecy.

Michael Weill, director of the Center for Information Systems Research at MIT, wrote in *Context* magazine about the head of a supermarket chain with US$5 billion annual turnover who spoke to MBA students at a leading business school. He spoke of the chain's innovative use of technology, including a "digital supermarket" pilot program that allowed customers to shop from home and have groceries delivered.

Afterward, one student innocently asked, "How do you determine the payoff of your firm's information technology investments?"

Borrowing a famous line about advertising, the chairman said: "I'm certain half of our information-technology investment doesn't pay off—the problem is I don't know which half."

"Technologists have developed capabilities that strategists have never figured out how to use."
— Technology Brief no. 1

So who's to blame for the cult of "return on Internet"? It's the same people who have created the cycle of false promises about measurable gains from technology: management consultants. Return on Internet is essentially a scam foisted upon the world by the promoters of Java and the Big Five consulting firms.

Traditional management consultants are looming large in the post-tech wreck landscape, claiming ebusiness as another string to their bow. Companies like Accenture, McKinsey, the Boston Consulting Group and IBM Global Services have spent the past year or so re-defining themselves as ebusiness consultants, offering the best of traditional consulting married with world-class online consulting. Trust us for your ebusiness consulting needs, they say, because we'll still be around in five years.

As they demonstrated through implementations of management information systems, SAP and ERP, they're technology experts. However, that focus on technology has come at the expense of the customer. They are constantly being led by what technology can do, rather than by customer need.

The result is investments that never deliver – projects that never finish. And because of the business model adopted by the Big Five, projects that never finish mean bills that never end.

Management consultants, most of whom have their origins in large accounting firms, traditionally enter open-ended, no-risk (for them), pay-by-the-hour contracts. It's like getting in a taxi and leaving the meter running, except you don't know where you're going and how long it will take to get there. The taxi driver is going to go the route he chooses to go and keep the meter running until you decide you want to get out, regardless of whether you're happy with the destination.

That approach just doesn't work in ebusiness – at least not for the client.

Clients need to ask themselves, "What am I getting from this solution?" It's not about the technology; it's about the outcome. The focus needs to be on delivering value to customers in a meaningful way

> *"If you can't explain it to your mother or your grandmother, don't do it."*
> **– Technology Brief no. 18**

Ebusiness developers, because they had their origin as small, passionate users of technology to solve a big problem, have been willing to stick their necks out and offer fixed fees that depend on whether or not the project is a success.

In ebusiness consulting, you need to enter an agreement of trust with your client – trust that you won't finish until everyone's happy. You can't just do a job as quoted and walk away, pocketing the money for time spent. You have to come up with a solution, but be willing to embrace uncertainty and the possibility that circumstances may change from the beginning to the end of the project, and not leave until it's working to the client's satisfaction. And satisfaction doesn't mean throwing money at extra consultants and handing the client the bill.

Ebusiness consulting is not about giving a man a very expensive fish. It involves teaching clients to fish in online waters. It means crafting a practical, flexible framework, and sticking with a project until the client's outcomes are achieved. Ebusiness consultants position themselves as partners, and do whatever it takes to meet their clients' needs.

Web-based online solutions are still technology solutions, but they have a much stronger focus on the person at the end of the chain – the customer. Developing online solutions requires a different way of thinking, a different mindset, than traditional IT-based solutions as practised by management consultants.

Management consultants tend to be very process-oriented – lots of reports, lots of measurement, no tolerance for deviating from the agreed plan, lots of testing to iron out the bugs before real people can use a system. If you come up with a better way of doing things halfway through an implementation, you'll have to endure the long and painful process of filing a change request and re-scoping and re-costing the entire project.

> *"Just because we haven't figured out how to monetize the technological advances doesn't mean that they aren't critically important."*
> — **Technology Brief no. 4**

The Web development process is focused more on outcomes and less on the specific steps to reach that outcome. It starts with a lot of questions: What do you want to achieve? What do your customers want? Do you really understand what you want?

It reflects the constantly evolving nature of the Internet. If the goalposts change halfway through, you just adjust the project, you don't force the client to stick with something that was relevant six months ago but may not be relevant today.

As *Fast Company* editor John Ellis says, "The traditional consulting model no longer applies. It isn't even close to the new reality of the Web. The traditional model said, 'We make a trade: your money in exchange for our smarts.' The Web model says, 'We collaborate – and get smarter together as we go along."

Thornton May from Guardent Inc. says, "If we want to get excited about technology again, we have to change the way we keep score. And generally approved accounting procedures shouldn't serve as the scorecard. The real scorecard is what matters to the customer, and technology isn't the main issue. It's the politics."

If someone offers to lend you a hand working out how much money you'll make bringing your processes online, make sure his other hand isn't already in your pocket.

ONLINE SUCCESS IS STILL HARD TO MEASURE

(PUBLISHED IN 2002)

The corporate website has well and truly moved past the stage of "We have a website because it's a good thing to do and it will pay for itself – eventually."

There are plenty of stories being written these days about the importance of achieving a return on investment (ROI) on your website and integrating your website with the rest of your business.

But when you look for specific information about actually measuring your website ROI, it's hard to find anything. That's because, despite all the user data that can be gathered by today's website measurement software, it's bloody hard to put that data into a useful context.

It's not for want of trying. The latest Web measurement software offers more reports, more raw data and more fodder for analysis than ever before. Many international consulting firms are now making a living out of ecommerce analysis.

In a landmark study, two of those consultants – NetGenesis and Target Marketing –conducted extensive interviews with senior executives from leading US websites to uncover best practices in ebusiness measurement.

Matt Cutler and Jim Sterne, the study's authors, were among the first experts to acknowledge that at its most basic level, ebusiness spans all major business processes, requiring businesses to integrate their ebusiness operations with their entire enterprise technical infrastructure.

They came up with classic lines such as, "The traditional management adage is 'You cannot manage what you do not measure.' The ebusiness addendum is 'You cannot measure what you do not define.'"

On the issue of defining ebusiness measurements, they said, "A large company without a standard definition for visits or users is likely to be using many different techniques for arriving at these measures. When the results are brought together for comparative purposes, it becomes especially difficult to tell what is really going on."

However, that study was done in 2000. When you visit the sub-website NetGenesis set up as a companion piece to that study to find out what they've produced since then, the answer is, not much. Two years later, they haven't performed a follow-up study or added new data to the original study. That indicates that even consultants specialising in web measurement analysis don't have new ideas on how to turn raw user data into ROI information.

That's why it's good to see an Australian book come out about website measurement. Hurol Inan, an ebusiness consultant formerly with Andersen Consulting and Deloitte Touche Tohmatsu, has recently published *Measuring the Success of Your Website*, in which he tries to solve this dilemma.

His insights include making the important point that when determining whether a website is a success, you can't limit your measures to website traffic. He writes:

"A mistake that may people still make in measuring the success of their websites is correlating its success with basic metrics – such as unique sessions served and page-views displayed. In reality, organizations should not build sites with the aim of increasing unique sessions and page-views. Rather, their website should focus on serving their customers and achieving returns for investors. Not only are these basic metrics meaningless, but also they can lead to incorrect interpretations, bad decisions, or no decisions at all. They cannot tell you if your online initiative is successful or not."

"Information gleaned from a measurement firm might show that average total time spent on your website is lower than that of your competition. However, this information might, in fact, mean that your website employs a better information architecture, making it easy for users so they don't need to spend as much time on it."

He warns against applying traditional advertising-type measures to online ventures.

"The temptation to use advertising metrics to measure the success of ebusiness is understandable, given the key role that advertising plays in the traditional economy. However, these standard advertising metrics are not fully transportable to ebusiness…. to get a complete picture, there is a need to measure all components of online activity. The metrics are designed to measure advertising events or campaigns, which are predominantly one-off or short-lived activities. Ebusiness, on the other hand, is ongoing and requires continuous tracking and measurement."

A strong customer focus is the key to success online, Inan writes. "The framework used to measure the success of online initiatives should recognise customer-centricity, and should help organizations to answer, in a practical and useful way, who your customers are, what are their needs and what sort of online behaviours they exhibit."

"Organisations must understand how the online and offline channels interrelate and support each other during the engagement of customers. This requires an identification of when and where cross-channel travels occur, and the integration of all interactions from multiple channels for an overall assessment."

He predicts that "Most organizations will seek ways to integrate user-activity data with other data sources to produce meaningful metric results and to perform detailed analysis."

Inan devotes a large part of his book to discussing the importance of quantifying the cost of converting a browser into a loyal customer, a theme echoed by Bryan Eisenberg and Jim Shreve from Future Now in their *Guide to Web Analytics*. They write that, "When it comes to online business, the metric that matters is the conversion rate. Other metrics matter too, but they combine together to make up the conversion rate."

The conversion rate measures the number of visitors who log on to a site within a given period divided into the number of visitors who take action there, such as buying something or registering their details.

But converting web visitors into web customers is only one role of a website. There are a lot of intangibles about ebusiness websites, things that don't necessarily translate into online transactions.

I know it's unfashionable to say you built your website because all your competitors have one and you don't want to be seen to be in the Dark Ages, but the fact is that not having a website has negative connotations for a business, regardless of how simple or sophisticated its web presence might be.

That feeling of "completeness" that a website brings, the knowledge that someone can touch your company and gain information 24x7x365, doesn't necessarily translate into sales on the website, but unmeasurably adds to your overall profit.

Another thing that is still too hard to measure is the value of investing today in expensive, complicated technology because you'll be higher up the curve in a few years' time and hopefully have the edge on your competitors.

And let's not underestimate the value of tyre kicking. Gathering information from a website without buying anything looks like wasted time in terms of online conversion rates, but it can result in stronger offline sales. Existing e-metrics have not yet come up with a simple way of measuring those activities.

I know at the beginning of this article I said we'd moved past the idea of having a website just because it's a good idea. But at the end of the day, in the absence of useful measurements, much of what drives corporate website development is the belief that it's a good idea to explore new ways of doing things. And that's not a bad thing. As the 18th century German philosopher G.C. Lichtenberg wrote: "I cannot say whether things will get better if we change; what I can say is they must change if they are to get better."

RETURN ON INSTINCT AS IMPORTANT AS RETURN ON INVESTMENT

(PUBLISHED IN 2004)

"Not everything that can be counted counts, and not everything that counts can be counted."
– Albert Einstein

While everyone agrees that websites are important contributors to a business' success, progress in measuring the size of the contribution has been painfully slow.

In his book *Web Metrics: Proven Methods for Measuring Web Site Success*, Jim Sterne recounts some of the recent history of web metrics. One of the first serious pieces of research done in this area was a series of interviews conducted by Sterne and Matt Cutler, cofounder of the business analytics company NetGenesis in 1999, which was turned into the white paper "E-Metrics: Business Metrics for the New Economy".

In their interviews, Sterne and Cutler found that "everybody was collecting huge volumes of information and nobody had a clue what to do with it. We learned that expectation and reality were not quite on speaking terms."

In 2001 Sterne decided to do another round of interviews "to see if we'd learned anything. This time.... I wanted to find out how people were *feeling* about Web metrics. How they were *coping* with the problem." What he found was that "everybody I interviewed was (and still is) begging for more information – not more data, but more usable information."

At about the same time Sterne was writing his book in 2002, I wrote an article examining what was happening in this area and stated, "There are plenty of stories being written these days about the importance of achieving a return on investment (ROI) on your website and integrating your website with the rest of your business. But when you look for specific information about actually measuring your website ROI, it's hard to find anything. That's because, despite all the user data that can be gathered by today's website measurement software, it's incredibly hard to put that data into a useful context."

Well, more than two years have elapsed since I wrote those words (which equates to 14 Internet years) and it's still incredibly hard. No widely-accepted method of website success measurement has yet been developed. There are more confirmed sightings of Elvis than there are websites that have measurably proved their worth to their company.

That's particularly true in Australia, where practical application of website measurement appears to be non-existent. An Australian study undertaken at the Macquarie Graduate School of Management (MGSM) this year (Editor's note: This study formed part of Ray's doctorial thesis on measures of success for corporate websites) showed that while websites have become an essential part of the corporate marketing mix in a relatively short period of time, that's due to a generalised belief that an Internet presence is good for business, rather than any strong measurable evidence of how much value a website creates for a company.

The researchers concluded that it is much easier to measure Web site productivity than profitability. Traditional accounting-based measures "can provide only a limited reflection of true shareholder value generated by IT investments" such as websites.

If you sell directly through your website, there is an obvious measurable benefit to a business. But most business websites these days are partly or wholly aimed at providing information and bolstering a company's brand image, not selling things directly on the site, and it's much more difficult to measure the success/ROI of non-transactional sites.

The MGSM researchers interviewed website managers at a range of Australian and international companies with a wide spread of types and sizes of business and turnover. They found that, 10 years after the birth of the Internet, one of the key reasons for having a website is still to "keep up with the Joneses" – in other words, most companies have a website because their competition has one.

The dominant attitude by business appears to be "we need to be on the Internet" without clear goals or research to back up the decision. Another Australian website study that examined websites for a group of West Australian car dealers contained this typical quote: "Jeez, we've got to have a bloody Web site now before everybody else out there does it."

The researchers found that "many companies are struggling to predict the likely impact of the Internet on their marketing and are wondering what they should do and how they should go about doing it, while those who move forward do so with mixed success".

They speculated that, as with many corporate IT projects, "the ultimate goal of corporate Internet projects may be simply to satisfy the company's board of directors and shareholders."

The study found that with the exception of the largest websites, measurement is *ad hoc*, and no sites surveyed had measures in place that tied non-tangible benefits in with standard measures. Those companies that were attempting to measure intangibles found it very complicated and expensive, making it questionable as to whether the money spent on the measurement effort was worthwhile.

Not surprisingly, the measure which companies said most often was important to website success was site traffic, particularly the number of visitors. Only the largest sites surveyed, however, had firm goals in place for site traffic; in most cases, if the trend was upwards, that was enough.

Interestingly, even though none of the companies interviewed was able to estimate the dollar value/ROI of their website to their company, most said that if their Web site was closed, it would result in a "significant loss in business" – and the bigger the company, the more likely the loss was considered to be significant. Several companies said the cost of doing business would go up, because they would have to go back to catalogues and there would be a corresponding increase in call centre activity.

Marketers responsible for Web sites are caught in a bind; because so much data can be collected via the site, there is an expectation (particularly from senior management) that it is easier to measure the impact of a Web site on the business than other marketing and sales activities.

However, because of the disconnect between online activity and offline sales (particularly for non-transactional Web sites), it's often more difficult, not easier, to measure the impact. The positive impact on a business may come in forms other than direct marketing benefits and increased sales, such as reduced costs, cost-effective added services or enhanced information.

The question a business needs to ask itself developing its website is, "What do we want a customer to do as a result of visiting the site?" While the ultimate answer to that question is "Buy more of our product/service", there are many steps that lead to that point, and that's where the value in a website most likely resides. It might be simply to be impressed with your service so next time they walk past your store they are more likely to enter, or to visit a competitor's website, so they will find out that you have the best product at the best price.

It's still incredibly hard to put a measurable value on those activities. But it may be enough just to know that value is there. After all, it's hard to measure the value of an ad placed in a traditional medium, but no one in business would say advertising is a waste of money.

AND THE GEEKS SHALL INHERIT THE EARTH

(PUBLISHED IN 2007)

Digital technology is ruining the party for creative types. Mind you, the rot started long before that. I blame people like Henry Ford, who perfected the use of assembly lines to build uniform copies of his Model T cars 100 years ago.

Ford was the Bill Gates of his time, driving (pun unintended) a technological trend into a society-changing phenomenon. His famous line, "They (customers) can have any colour they want, as long as it's black" was one of the first examples of science (and commerce) triumphing over art.

185

That's not a bad thing – can you imagine what cars would cost if they were all hand-crafted? – but it has been a trend that has moved into most other areas of business. Think property developers and their cookie-cutter houses, IBM and computers, Microsoft and software, record companies and sausage-factory music, and film companies and blockbusters on thousands of movie screens. One notable exception to this is Apple – Steve Jobs seems to be able to hold both creativity and mass production in his hands at the same time.

And now it's happening to the last bastion of creativity – the advertising industry. Advertising has steadfastly held out against the scientific push for more than 100 years (as a contemporary of Henry Ford, US department store merchant John Wanamaker, once said, "Half the money I spend on advertising is wasted; the trouble is, I don't know which half"), but that resolve is now crumbling after more than 10 years of ad measurement development on the Internet.

The ultimate example of this is the recent purchase by global advertising behemoth WPP of 24/7 Real Media for US$650 million. WPP is the world's second-largest marketing services company and owner of media agencies such as MediaCom and MindShare, as well as creative agencies Grey Worldwide, JWT, Ogilvy & Mather and Young & Rubicam, as well as several interactive specialist agencies.

24/7 Real Media, meanwhile, is a marketing technology and digital advertising measurement company that, as it says on its website, "brings the science of digital marketing to advertisers and publishers around the world."

WPP's chief executive Martin Sorrell said, "We focus on creativity and media. This adds a futher dimension. It's the application of science to our business. We think this is increasingly important."

WPP's stated goal is to draw as much as a third of its revenues from digital sources, and this acquisition is going to accelerate that process.

Real Media founder Dave Morgan (who sold his interest in the company several years ago) told the media, "This is a strong, forward-thinking move for WPP. Instead of buying Digitas [WPP's rival Publicis bought the interactive agency earlier this year] or another ad agency, WPP is gaining the upper hand in ad serving so it won't have to defer to Google or aQuantive in the future."

Microsoft's acquisition of aQuantive, announced in May, followed Google's announcement that it would pay $US3.1 billion for online advertising technology company DoubleClick. Microsoft was reportedly interested in DoubleClick as well and it's believed aQuantive's ad serving technology could help the world's largest software manufacturer close the gap with Google.

aQuantive doesn't have the same public profile as DoubleClick, but it boasts a suite of quality companies and products, including online ROI toolmaker Atlas, the DRIVEpm ad service and Razorfish, one of the largest interactive ad agencies in the world. Although the price Microsoft paid was stratospheric (the cost per impression was more than twice what Google paid for DoubleClick), it's shaping up as a quality purchase for the company.

Technology and finance blogger Paul Kedrosky wrote that "the towering premium Microsoft was willing to pay for the company (is) Murdoch-ian, the kind of 'shut up and sing' (or at least shut up and sell ads) price that says let's stop talking and just get this over with." Yahoo!, meanwhile, has also gotten in on the act by taking a strategic stake in Right Media, which auctions online ads in real-time to the highest bidder, with more than 2 billion impressions traded daily.

And almost as if to counterpunch Microsoft for moving into its territory with the aQuantive purchase, Google bought online security solutions provider Postini in early July for US$625 million. Postini will make it easier to persuade larger businesses to make the switch to Google

Apps, which is competing head-on with Microsoft Office. Google claims more than 1,000 small businesses a day are signing up for Google Apps, but it is not making headway with medium and large businesses.

What do all these boardroom shufflings mean? F. Ahrens, writing in the *Washington Post*, summed it up well: "The art of advertising is turning into the science of advertising. Agencies now need math guys." He quoted interactive agency chief Bob Greenberg, who said, "Technologists are pretty foreign to the traditional agency model, but they're an important part of the future. Traditional creative is becoming less and less important."

Does this spell the end of creativity in advertising? Of course not. But it does mean accountability for results is becoming more and more important as developments in online technology increase the availability of realistic measures of the impact of advertising and marketing.

The Cannes Festival will always be there as a monument to creativity for creativity's sake, but as more agencies and software companies integrate digital serving and measurement into their organizational DNA, more and more energy will be focused on achieving commercial outcomes – which, after all, is the purpose of advertising and marketing. Get used to having more "maths guys" in your planning meetings; Learn to understand their language.

MOBILE

IPAD: FAD OR FUTURE?

(PUBLISHED IN 2010)

Kenneth Boulding, curmudgeonly scholar, peace activist and founder of the evolutionary economics movement, died in 1993, but if you look at his quotable quotes today, you'd swear he was talking about the iPad, introduced nearly 20 years after his death.

Apple moved 250,000 iPads on the weekend of its release back in April, and since then it has racked up sales of well over 10 million units. This has led to the bizarre situation of analysts downgrading Apple stock because it may sell 'only' five million iPads in the last quarter of the year, instead of the wildly optimistic six million predicted a couple of months ago.

Which leads me back to Boulding. He said, "If you believe exponential growth can go on forever in a finite world, you're either a madman or an economist." I won't comment on which category I place the analysts who are taking Apple to task for not putting an iPad in the hands of every man, woman and child in the Western world in the space of 12 months.

TOOLS SHAPING US

Boulding was also famous for saying, "We make our tools, and then they shape us."* Again, he could have been writing about the iPad and its burgeoning list of competitors, which have been enthusiastically embraced by everyone from dead tree publishers to health care conglomerates during 2010.

There's been so much buzz about how this 'revolutionary' concept will change the way people interact with technology. But I've been around long enough to remember that tablet computers have been around for yonks. Remember the Palm and the Newton (made by Apple, no less)? As MacWorld said about the iPad earlier this year, "It's a product in a category – tablet computing – that has been a flop despite a decade of hype."

So, what has made 2010 and the launch of the iPad different? I posed that question to Peter O'Neill, founder and CEO of 2moro mobile, during a recent podcast. He said the key to the iPad's success lay not with the device itself, but its predecessor, the iPhone. "The release of the iPhone in 2007 was a real game-changer," he said.

For several decades, mobile phones improved in incremental steps – smaller and smaller (and then bigger), more ergonomic keypads, more memory, cameras, etc. etc.

By developing a user-friendly touchscreen and emphasising everything but making phone calls, the iPhone turned the idea of a phone sideways and changed the way people behave.

In the words of WIRED's Fred Vogelstein, "This 4.8-ounce sliver of glass and aluminum is an explosive device that has forever changed the mobile-phone business, wresting power from carriers and giving it to manufacturers, developers, and consumers…. Now, in the pursuit of an Apple-like contract, every manufacturer is racing to create a phone that consumers will love, instead of one that the carriers approve of."

Particularly in Australia, the iPhone is de riguer for anyone who claims to work in the digital space – though this could backfire among those of us who rail against group think.

Peter O'Neill says the iPhone was launched at just the right time in the Australian market and have made it particularly hard for other entrants. He points out that Google Android phones are currently outselling iPhones in the US (there are, admittedly, more models available), and Blackberries have a large and loyal audience. But in Australia the iPhone's premium pricing doesn't seem to deter its take-up.

SOME THINGS GO BETTER WITH APPS...

Peter argues out that the iPhone paved the way for thinking differently about tablet computing, and the iPad and its competitors are reaping the benefits.

The iPad has been sucking up most of the oxygen in this area. If you search for the word iPad in Google, you get more than 207 million results – not bad for a word that didn't exist 12 months ago.

Many apps that work on the iPhone also work on the iPad – or have been easily converted – so there are already more than 100,000 apps available for the device, giving them a leg up on competitors.

So why should marketers get excited about this? Peter O'Neill says that "Some things can only be done as an app. Enterprise apps – for example, property inspections, surveyor tools – can overtake paper." Field workers can collect information offline and synch with their computer when they get back to the office.

Book and magazine content are obvious choices for the iPad. Peter also points out that the iPad is a "fantastic babysitter. The content on the iPad is better than the iPhone, kids love it, and importantly, an iPad is not as critical a tool as phone."

...AND SOME THINGS DON'T

The biggest risk in developing an iPad app, according to Peter O'Neill, is the Apple review process. "If you have tight deadlines (for the release of your app) it could be a problem – dealing with Apple can be scary and frustrating." Google's equivalent of the App Store makes it much more flexible and easier to make apps available for downloading.

Letting your target audience know that you have an app is another issue. In the early days of the iPhone, there were a number of success stories of organic growth. Now, you have to market your apps through other means, such as social media, newspapers, TV, and your corporate website.

And of course, there is the Flash problem. Peter says using HTML 5 will solve most, if not all, of the problems associated with optimizing websites and apps for the iPad. It "adds a level of complexity," he says, but if you avoid flash and your app or mobile website works on most browsers, it should work on the iPad.

He said the mobile Internet device market – which covers both smartphones and tablet computers – is "like the Web was 10 years ago. It will only get more complicated, not easier." So, if you're a marketer, now's the time to get to grips with it.

* Another of my favourite Boulding quotes which has application today, although nothing to do with the iPad, is "Economists are like computers. They need to have facts punched into them."

WHY MOBILE MATTERS

(PUBLISHED IN 2011)

What a difference a decade makes. Ten years ago, when the tech world was still picking up the pieces from the broken bubble, those of us optimists knew that there was a strong future ahead, but we couldn't say what that future would look like.

Ten years ago, it was believed Internet uptake was well established, with nearly 50% of homes and most businesses being connected to the Web. Ecommerce and video on the Internet were identified as growth areas, but it was hard to imagine how all that could be achieved.

Today, 80% of adults use the Internet, which is pretty close to saturation and a good solid figure. But the real story about Internet use, in Australia and around the world, has been the way people are getting connected to the Internet.

Whereas 10 years ago being connected to the Internet meant a desktop or laptop connected via a blue cable to the copper telephone network, today it means something a lot more flexible and powerful.

Let's look at a few numbers. The proportion of households with broadband has increased from 5% to 90% since 2000. Mobile phone ownership has increased from 50% of the population to 82% of the population – and most of those phone users are accessing the Internet via their phone, a concept unthinkable to all but the most die-hard Dick Tracy fans 10 years ago.

Wireless connection to the Internet wasn't even available 10 years ago; today nearly 60% of us connect wirelessly, either via a home wireless network or via a mobile device (phone or tablet).

Meanwhile, online social networking, which didn't even exist in 2000 (unless you count Geocities* – remember that?), is now a tool used regularly by nearly 50% of the population in places like Australia and the US. According to Nielsen, Australians are the biggest users of social media in the world, clocking up nearly seven hours per month on Facebook, Twitter, etc. Nearly 75% of Australians have used digital social media at least once.

Storing data on outside servers has also exploded. "The cloud", a phrase that didn't even exist 10 years ago, is now being used by more than 66% of people.

THE GREAT INTERNET SHIFT

The more than 17 million Australians who are online spend 22 hours online per week online, compared to 15 hours watching TV and 3.4 hours reading newspapers, according to Nielsen stats. Online use has exploded, while TV use has stayed pretty steady.

So, if we're not watching less TV, where do we find the time to spend so many hours online? Because 40% of us use the Internet at the same time as we're watching TV.

A few years ago this would have meant either having your desktop in the same room as the TV or sitting in front of the box with your laptop on your lap. Today, more often than not online/TV multitasking means sitting in front of the TV while you're on your phone or your tablet.

I can relate to that. Although I could pick up my laptop, carry it out to the family room and sit upright typing away on the device while watching TV, I have rarely ever done that. However, while I'm watching the box, my iPad stays within arms' reach and gets used without fail every night. My kids, meanwhile, are constantly using their phone in front of the TV (don't get the wrong idea – we talk to each other, as well!).

RISE OF THE MOBILE WEB

Fuelled by the development of the Apple iPhone and rocketed along by its various incarnations, along with competing phones from RIM (Blackberry) and Android, Australians have embraced smartphones faster than just about any other nation, with mobile web adoption growing at a rate eight times faster than the desktop web. At the end of last year, 12% of all Google queries were coming from mobile devices, a figure that has grown 3000% in the past three years.

This has meant a dramatic change in when and how we use the Internet. While desktop Internet use peaks in the middle of the day, mobile web use peaks at 11 at night (I'm not going to speculate on what that says about what Australians do in bed these days!). Mobile web use also peaks on the weekend, the same time where desktop-based web use hits its lowest point.

These days, talking to people and sending text messages only accounts for 32% of the time we spend on a mobile phone. The rest is made up of email, using web apps, playing games, checking Facebook, watching videos and searching for information.

I could drag up plenty more statistics, but my point is that not only has Internet use skyrocketed but just as importantly, the way we use the Internet has shifted dramatically over the past decade. The era of mobile has well and truly arrived.

The desktop-based Internet is not dead, but the use of the Internet via mobile devices has just exploded, and businesses need to come to terms with this development.

You can no longer just consider how your website looks on a desktop/ laptop monitor. In 2011 you have to be available on all devices and it all needs to work. It's no different to being on a supermarket shelf.

The web just has to work – usability is key. And when I say web, that means on a desktop, a laptop, a smartphone, a tablet – all of the above.

What do your customers see when they access you from the mobile web? Do you even know? It's high time to find out.

*As an aside, amid all the criticism of News Corporation's business judgment for paying US$580 million for MySpace and selling it recently for less than US$50 million, it's worth noting that Yahoo! paid US$3.5 billion in stock for Geocities at the height of the tech stock boom in 1999, and closed it down completely in 2009 – now that's a write-off!

APP OR FAP? THE DEBATE OVER APPS VS. MOBILE WEB

(PUBLISHED IN 2011)

Following on from my last post, mobile is becoming increasingly critical to your business' web strategy.

So what's the best way to tackle the mobile web? Is it just a matter of designing a one-size-fits-all site that looks good on a small screen as well as a large one? Do you just forget about a mobile site and go for an 'app'? Or is there a middle path you can take?

Some definitions are useful here, as this is a quite confusing area. In computer terms, an application is a piece of software that performs a task for you. By that broad definition, programs like Word, Excel and PowerPoint are considered to be applications.

With the appearance of the smartphone, which is as much a computer as a telephone, applications have been developed that allow you to perform tasks on your phone. Maybe in an effort to distinguish these applications from software used on desktops and laptops, they tend to be called apps for short.

You buy an app (or download it for free) from an app store, and then you own it and can use it over and over to perform that above-mentioned task.

This is distinct from the web, which you browse through, visiting websites that may or may not let you perform a task.

As tech blogger Daniel Odio writes: "Being on the web is a very nomadic experience. You visit your favorite sites, then you visit other sites, but there's very little sense of ownership on the web. An app, by contrast, provides a unit of ownership. You purchase an app and you own that contained experience."

Or, as another blogger Bill French puts it: "Apps have become a meaningful abbreviation to technology that just works. Apps provide a common and easily understood idea that has been widely accepted as a solution – indeed a means to get stuff done quickly and effectively."

When you say 'app', most people think of an iPhone app that you download from iTunes. While this is the most popular form of app – more than 350,000 apps are available on iTunes, with more than 15 billion downloads – these days there are also tens of thousands of apps for Android phones and Blackberries, as well.

An important thing to realize is that an iPhone app won't work on an Android phone, and vice versa – it's like the good old days of Mac and PC where programs would only work on certain brands of computers.

In fact, it's worse than that; as well as Apple, Android and Blackberry, there are more than 6000 different types of phones on the market, so if you develop an app for one type of phone, there is a long, long tail of devices that your app won't be compatible with.

MOBILE WEB APPLICATIONS – NOT THE SAME THING

As an alternative to device-specific apps, technology companies have developed mobile web applications, which have similar functionality to apps, but which are served from a website rather than being downloaded onto a mobile device.

They can look and feel much the same as an app – you can even download a shortcut to your desktop that you click on to 'launch' the application, the same as you would with a device-specific app.

While they can perform many tasks, mobile web applications aren't quite as flexible or quick as apps. However, they have the huge advantage of being able to be used on many different types of phones or tablets.

Here's a summary of the advantages and disadvantages of apps vs. mobile web apps:

APPS – THE GOOD

- They start fast and can usually be configured to run in the background
- They can work without accessing the Internet (so you don't have to incur international roaming charges while you're travelling overseas).
- There is a wide range of functionality that can be built in to them.
- They can authenticate a user's device with a password – useful if you leave your phone in the back of a cab.
- They can be used for credit card payments.
- They can access phone hardware, such as an accelerometer or GPS
- They can be used to push messages on demand from the network.

APPS- THE BAD AND THE UGLY

- They are device-specific, which means you have to basically start from scratch when you want to convert your iPhone app for use on an Android, or a Blackberry, or one of the 6000 other phones on the market.
- The user's app data is vulnerable when the phone is lost/stolen/compromised.
- You can't recall, delete or secure what has been published on a specific device.
- It can be expensive and time-consuming to get your app published in the iTunes store.
- You have to continually download updates to the application to keep it up to date.

MOBILE WEB APPLICATIONS – THE GOOD

- They can work on a variety of devices
- Like apps, they can be used for credit card payments

- You don't have to download an application or any maintenance updates; every time you open the application, it automatically delivers the most up-to-date version of the application
- You can bookmark the application to operate just like an app on the desktop
- If you just want a one-off interaction with an app, you can receive immediate access without a download
- They are much cheaper and faster to develop and maintain.
- Direct billing and version updates for subscription-based content are much easier launched via a web browser

MOBILE WEB APPLICATIONS – THE BAD AND THE UGLY

- They are slower to launch than apps
- Their functionality can be more limited than apps
- You can't find them in an app store when you're searching through a list of apps (though this can also be seen as an advantage.

WHICH WAY SHOULD YOU GO?

So, given a choice, should you build an app, or a mobile web application? At the moment, it's not an either/or situation.

What I would warn against is the "let's build an iPhone/iPad app" mentality that exists at the moment, fuelled by the buzz around Apple products in Australia. That narrow way of thinking is not justified by the restrictions you impose on yourself if you go straight to building an app.

Don't lay awake at night thinking about ways to use an app for your business. If you have to do that, you shouldn't be using an app.

You really need to weigh up the pros and cons of the specific situation you find yourself in. What is it you're trying to achieve with your application? Who is your target audience? Functionality, not hype, should drive your decision.

THE CASE FOR MOBILE WEB APPLICATIONS

(PUBLISHED IN 2011)

I am a proud Apple user and rely on several devices powered by iOS in my professional and personal life. I'm not alone – iPhones make up 45% of smartphones sold in Australia, one of the highest market shares in the world.

But the mobile world is changing rapidly, even more rapidly than other market sectors. Despite the continuing success of Apple, you cannot discount the presence of its competitors, most notably Google.

In the US, for example, Android phones (powered by Google) now have a 40% market share, almost 15% ahead of the iPhone, according to the most recent Comscore report. Meanwhile, according to Canalys, Android's global mobile phone market share is 48%, compared to only 19% for the iPhone.

And that's not to say Android will gain total market dominance. Samsung and Nokia have substantial shares, and the Microsoft Windows phone will have a significant impact in the market over the next few years.

My point is that I believe that if you develop device- or operating system-specific applications, you are severely limiting the usefulness of those applications. If your applications are aimed at a wide audience, device-specific apps restrict the utility of your tool to users of one type of phone – or they double, triple, quadruple and more the cost of your development if you want them to be available more widely.

In Australia, there are more than 1200 types of devices that access mobile websites. In the US, that figure is more than 2200.

Even in Australia today, if you develop an iPhone app, you will reach 45% of smartphone users, but you also need to connect with the other 55 percent. The simplest, most cost-effective way to reach a wide mobile audience is through mobile web applications.

In a recent podcast I spoke with Lucas Challamel from NetBiscuits, a company that provides cross-platform mobile websites and applications.

He says there are three major challenges for businesses who want to bring Internet content to mobile successfully and profitably:

- Dealing with the fragmentation mentality associated with software and mobile development
- Creating a useful service
- Working out how to make money out of that service

NetBiscuits, a German-based company now operating around the globe, including Australia, was developed do deal with the fragmentation problems associated with the development of device-specific apps, or as he calls them, native apps.

He says the comparison between mobile web applications and apps is like a racehorse vs. a showhorse. "Mobile sites are a requirement for today and into the future, while apps will remain a preference," he says.

The list of devices which Netbiscuits tests to make sure its mobile web applications have nearly universal coverage numbers more than 6000 devices, and it grows by more than 30 per week.

Without delving too deeply into a technical discussion, much of Netbiscuits work revolves around working with web standards such as HTML5, which are being developed to make it easier to offer websites and mobile applications that work the same across a range of devices. Lucas Challamel calls HTML5 a "standard in motion". "It may become the one power to rule them all, but the standard is still not finalized."

While waiting for that standard to be finalized, Netbiscuits has developed interim solutions such as 'hybrid apps', which are device-specific, downloadable apps that have some or all of their user interface in an embedded browser component.

Like a device-specific app, a hybrid app is downloaded from an app store or marketplace, it is stored on the device, and it is launched just like any other app. But to developers there is a huge difference, because instead of rewriting the app from scratch for each mobile operating systems, they write at least some of their application code in programs such as HTML, CSS and JavaScript, and reuse it across devices.

Lucas Challamel says that, "Fragmentation is here to stay; businesses need to cope with it."

In my view, a marketer's central task is to make the brand easy to buy and this requires ensuring people can find it and know about it. Everything else is secondary.

In a fragmented mobile market, developing a mobile web application which can be accessed by just about any mobile device with a web browser is, in my opinion, the most effective way to make your brand easy to find.

SEARCH

WHAT YOU REALLY NEED TO KNOW ABOUT SEARCH

(PUBLISHED IN 2007)

The art and science of search engine marketing is developing at a rapid pace. More and more customers are using the Internet as their first port of call in making a purchase decision and as a result, more and more companies are using paid and unpaid search placement to direct customers to their business.

As search engines like Google change their search algorithms daily to stop unfair manipulation of search terms, there is a growing and bewildering array of techniques that need to be mastered in order to stay ahead of competitors in the fight for relevant traffic.

Meta tags, keywords, image tags, niche directories, link building, pay-per-click, AdWords – those are just some of the simpler terms that are taking up more space in a marketer's brain.

And the amount of brain space required for search-related thinking grows every year. Search expert Greg Jarboe, one of the speakers at this year's Search Engine Strategies (SES) conference in New York, pointed out in his column in ClickZ that this year's conference included 33% more sessions than at the 2004 SES conference. Meanwhile, only about 10 sessions had the same title as sessions in 2004.

Jarboe writes, "This means that more than 80% of what we learned in 2004 (about search engine marketing) is no longer being taught in 2007. Or, to put it another way, less than 20% of what you need to know today is something that you could have learned three years ago."

So, what do marketers really need to know? The first thing is to be familiar with the new language and the new acronyms being thrown around. For example, you need to understand the difference between search engine optimization (SEO) and search engine marketing (SEM). You need to do both, but you need to get SEO right first. Once your site is optimized for searching by the major engines, then you tackle the marketing issues.

Successful SEO comes down to balancing the needs of three disparate groups – your company, the search engines and your customers. The main business of search companies such as Google, Yahoo and MSN is to sell their advertising and other business programs. If they have more users than their competition, they will be able to attract more advertisers and business partners.

That's why the search engines' main goal is to attract more users than other search engines by presenting users with the largest, best quality list of websites as quickly as possible when someone enters a keyword search phrase in the search box and hits submit. Symbiotically search engine users are most likely to use a search engine that gives them a quick list of quality web sites or information (quality meaning it meets their information need, whatever that was) to their search queries.

If a search engine gives them lousy or useless links, or takes a long time to load, they'll leave and use another search engine. Users will also stop following search links if they take them to web sites that are not interesting and provide poor content.

What are the most important things to get right in SEO? Seattle-based search marketing consultancy SEOmoz recently surveyed 37 global SEO experts and came up with the most important factors for top SEO results. They were, in order:

KEYWORD FACTORS

- Making sure the optimal targeted search term or phrase is placed in the title tag of the web page's HTML header
- Using the targeted search term in the visible, HTML text of the page
- Ensuring topical relevance of text on the page compared to targeted keywords

TOP PAGE ATTRIBUTES

- Making sure all links on a page point to high quality, topically-related pages, both internally and externally
- Finding the balance in the age of a document – older pages are seen as more authoritative, but younger pages may be more temporally relevant
- Securing the right amount of indexable content – how much visible HTML text is on a page

SITE/DOMAIN ATTRIBUTES

- Obtaining a high global link popularity of site
- The more older content available on a site, the better
- Ensuring that inbound links to a site are topically relevant

That's just scratching the surface of SEO – SEM then takes you to a whole other level of activities and reports. It is a metrics-driven, pragmatic marketing approach that is not nearly as intuitive or creative as pumping out TV ads. You could call it the revenge of the left-brainers.

If you're having trouble getting your head around this, you may decide you need professional help rather than trying to do this yourself. But even if you rely on external agencies to do your SEO and SEM, you will get better results out of your agency if you understand the landscape.

One way to understand this a bit better is to head out to one of the many search conferences that are being held in Australia and internationally. Search events are a burgeoning business, and not just because there is money to be made for conference organizers – businesses are hungry to learn more about search.

Search Engine Strategies is probably the highest-profile international conference, and they are holding 10 conferences this year in several countries beside the US. Although the speakers and topics may seem confusing and geeky, you'll no doubt meet many marketers who are also trying to understand this growing area of marketing.

While search marketing makes up nearly half of all online marketing media spend, it is still less than 8% of all media spend internationally – and that percentage is likely smaller in Australia. Today, that is – don't kid yourself by thinking this area will stay small or go away.

Start training your mind to look for small but lucrative opportunities through search. As Aaron Wall, author of the *Search Engine Optimization Book*, says, "The web is like New York City or pornography: there is so much demand, implied intent, and capital that even small niches can become highly profitable if you target your site and brand around them."

THE IMPORTANCE OF SEARCH ENGINE OPTIMISATION (SEO)

(PUBLISHED IN 2010)

Number one on Google – that's the Holy Grail that search engine optimisation (SEO) was meant to deliver to savvy site owners.

SEO involves tweaking your website to appear in the first page of Google search results for the keywords your customers are most likely to enter,

And yes, it's Google we care most about, because Google still reigns on the Internet.

According to search engine data collection agency ComScore, Google accounts for around 67% of all search queries worldwide. Next in line with just 8% is Yahoo, then Chinese search engine Baidu, which holds 7% of global search. The remaining 18 percent is distributed among a range of other search programs including Russian engine Yandex, Bing, AOL, Ask.com, eBay and Facebook.

In Australia the figures are even more stark. Recent Hitwise data shows that Google Australia accounts for nearly 75% of Australian searches, with Google.com taking more than 12%, equating to an overall market share of more than 87%.

In comparison, Yahoo and Yahoo Australia combined for only 7% of Australian searches, while Bing accounted for only 4% of the market.

But as search engine optimisation has moved from a contest of geek-stardom to a critical business feature – and more and more companies are vying for that page one spot – ranking highly in search is becoming far more challenging for most organisations.

Australian search expert Jeremy Tang, who I interviewed for a recent podcast, says that the biggest issue for most companies is a lack of understanding about how SEO works.

SEO is not just about rankings, he says; it's about optimising all of the organisation's communication (online and offline) to the customer, not to the search engine. "You need to understand exactly what your customer is looking for - and make sure you deliver that," he says.

"Get that right and you'll not only do well in Google ranking, but more importantly, your customers will be more engaged because you've been able to address their information needs better than your competitors."

Jeremy firmly believes that SEO has to be an ongoing process to be effective, because your customers' needs are always changing and the competitive environment is dynamic.

If you want page-rank nirvana, though, you have to make sure you get a combination of things right. "The most technically optimised website possible won't rank for terms that are important to your customers without the right content strategy," Jeremy says.

"Or you might have a great content strategy - but if much of your content isn't visible to the search engines you'll get sub-optimal results."

Google SEO spokesman, software engineer Matt Cutts, recently gave a presentation to search engine experts showing the results of a site audit that Google did on its own website.

Cutts is one of Google's most accessible personalities and search engine experts wait breathlessly on every tip he shares about getting good search engine results. (He is so geek-cool, he's even been immortalised as a cut-out and dress-up cartoon doll.)

Cutts gets peppered with questions about the secrets of Google's algorithms and how much difference tiny tweaks and links and strategies might make to the ranking of a website.

But a lot of the time, the advice he gives is not about mathematical algorithms or convoluted ways to raise your site's rankings – it's about thinking like a customer.

"Think about what users are going to type to find your page," Cutts told *USA Today*. "Then just make sure those words are on the page."

Jeremy Tang agrees. He says having good insight into what information your customers are looking for is a critical first step. Next, you need to deliver that information through your website in your customers' natural language, in a format that search engines can read. And finally, you need to build the authority of the website as a trusted source of this information through backlinks.

"The success of the program will hinge on that first step: understanding your customer sets the entire direction into which everything else falls into place," Jeremy says.

And these days, there are many ways of finding out exactly what your customer is looking for – even when they are not too sure about that themselves.

"Search is behaviour," says Jeremy. "So when you are trying to understand your customers, always pay more attention to what they do, than to what they say."

Plenty of companies base critical decisions about their web strategy on the opinions of management - or at best, using customers' self-reports from focus groups and surveys.

But it's possible now to tap into the actual search behaviour of millions of Australians to find out exactly what they are looking for.

"Not only is this statistically more significant as a data source – it is all based on actual behaviour," says Jeremy. "This leads to much better marketing decisions."

Some businesses make the mistake of believing that they can change the terminology in the marketplace – but the web is beyond their control.

On the web, people will search the way that makes sense to them. If your website doesn't include the critical keywords for your market, your customers won't find you.

Jeremy says that he has had several clients recently ignore recommendations for keywords, despite being presented with search data on the terminology people used for the product they were selling.

Both times, the client believed their multi-million dollar advertising campaign would dictate - or at least heavily influence - the terminology in the market. Both times, the client was wrong.

"These days, running TV, radio and print ads will trigger people to search to find out more. If they can't find the information in the language they are using to search for it, then their interest is lost - or worse, intercepted by a competitor who understands search better," says Jeremy.

If you are going to optimise your website so that it appears high in the search results for your keywords – you've got to be really sure of your keywords and research them well.

Keyword research is about understanding the conversations your customers are having; and the best way to capture this conversation is to ask the people in your business who interact with your customers every day - your call centres, your sales staff, customer service, resellers, and of course your actual customers.

"Having enough conversations with these stakeholders helps to quickly uncover patterns in what they are looking for," says Jeremy. "Team that up with online research by a qualified SEO person, and you'll get some powerful insights into customer interactions offline and online."

Once your customers find you – there's got to be good reason for them to stay on your site. For that reason, if people are searching for information that your brand is in the best position to provide and you aren't ranking for that search, someone else can control that messaging.

Building deep, continually refreshed content will build your brands reputation as a trusted source of information as it continually appears in the searches your customers are making.

If you want to develop a deep relationship with your customer, you have to provide them with information that they are interested in and keep it constantly updated because their needs are dynamic. "It makes sense from a human perspective which is why it works from a SEO perspective because Google is all about delivering relevant and valuable customer experience," says Jeremy.

TECHNOLOGY

THE X-INTERNET FILES, OR THE END OF THE WEB AS WE KNOW IT

(PUBLISHED CIRCA 2000)

> Interviewer: What's the Next Big Thing in online media?
> Kurt Andersen (author of "Turn of the Century" and founder of Inside.com): Plastic?
> **– From interview in "I Want Media" (www.iwantmedia.com)**

I remember back before most people had heard of the Internet (all of 10 years ago) when I first read about this concept called "hypertext". The article tried to explain how you could look at a photo of ducks on a pond on a computer screen, and by clicking on various parts of the picture (although they didn't use the term "click" back then; they called it "drilling down") you could find out all about various types of ducks, things made from duck down, who took that particular photo and other photos by the same photographer, even songs and poems that had the word "duck" in them.

It sounded exciting, grandiose and pretty near impossible to achieve. But within a couple of years it was a reality, embodied in a user-friendly offshoot of the Internet called the World Wide Web.

The Internet had already been around for more than 20 years at that stage, but it consisted of screeds of impenetrable computer code appearing in black on a grey screen, the domain of defence experts, academics and a handful of computer geeks. The Web, so-called because of the similarity of connected computers to a spider's web, presented the Internet in a way ordinary people could understand.

The rest, as they say, is history – the Web's spectacular rise during the 1990s and its equally spectacular fall since the turn of the millennium. Many analysts are now predicting the death of the Web. Could they be right?

Let's face it: the Web is far from perfect. It's great at some things, but not so good at others. The Web is at its best when it comes to news, information searching and community-building. But if an online activity is bandwidth-hungry, despite the emergence of the science of website usability, faster modems and more powerful computers, the World Wide Web continues to be the World Wide Wait.

The worst offenders when it comes to slow downloads are ecommerce sites. In a recent survey of consumer ecommerce websites, Forrester Research found on a scale of -50 to +50, where +25 represented a passing score, the average user experience score was -3.

With those sorts of scores, it's no surprise that analysts are predicting that because of "cumbersome user interfaces and static, bandwidth-hindered experiences" as labelled in E-Commerce Times, the Web's long-term value as the primary ecommerce platform is limited.

But when you think about it, websites were invented to share information, not security and transaction data. And that's why the next phase of development surrounding online transactions and ecommerce is happening outside of the Web. It's been called the X (for executable) Internet.

"The Web will be augmented and changed, not obsolete," according to Jupiter Media Metrix analyst Rob Leathern. "Technology companies will move away from the Web as a static published page, and will be more dynamic about how transactions and data move around."

Internet applications will execute code on the user's PC and other devices to enable one machine – or shopper – to talk to any other one on the Internet. Shoppers will be able to purchase items online in seconds, using familiar drag and drop desktop skills rather than waiting for pages of text and graphics to load.

In other words, forget Amazon and think Napster and instant messaging. The key to the success of the X Internet will be having key parts of ecommerce applications resident on the user's PC – or, as time goes on, their mobile, PDA, TV or even fridge.

Forrester Research is predicting that by 2005, shopping online will be an interactive, dynamic experience on the executable Internet, and by 2006, the number of non-personal-computer Internet devices – an 'extended' Internet – will outnumber PCs.

According to Gartner Research, by 2005, US$13.4 billion worth of b-to-c ecommerce transactions will be done with a TV, US$9.5 billion will be done through a mobile device, and US$204.8 billion will be generated via PCs. By that stage more than 40% of consumers will regularly use multiple platforms to make transactions, the study said.

One practical outcome of the X Internet will be what's being called "meta services", bits of user and transaction information shared among companies and other users to seamlessly conduct multiple related transactions at once. For example, when you're booking a business flight the Internet will automatically check whether your colleagues are available for a meeting, and offer you travel guides you can buy either online or at the airport.

This will also finally make the couch potato's dream become a reality: the Internet-enabled fridge that senses when you're low on beer and sends out for another slab.

But don't expect the X Internet to appear overnight. Major retailers and companies such as Amazon.com have made huge financial and behavioural investments in Web-based ecommerce, and they're not keen about chucking out their current models and starting over. As a result, analysts are predicting adoption of this new commerce paradigm will be slow. Also, the sharing of personal information involved in executable ecommerce is likely to raise privacy concerns.

So is it time to get in early and invest in Internet start-ups again in the hopes of making some fast money in the next Internet boom? Don't count on it. As John Landry, head of investment firm Lead Dog Ventures and former CTO at Lotus Development Corp said, "I doubt that X Internet will result in a boom like we saw with the Web in the late 1990s, because I don't think anybody is dumb enough to fall for that again, but it will set off another cycle of new and innovative development."

TORTOISE TIME FOR ONLINE TECHNOLOGY

(PUBLISHED IN 2002)

Two years on, the waves from the global technology shakeout are still being felt. In Australia, Open Telecommunications, the once-huge telco founded by Wayne Passlow with the millions he made from selling his Sausage Software shares, called in the administrators in

July. The headline in *Slattery's Internet Watch* said it all: "Another one bites the dust!"

The same day, web design group Spike, the company that at its peak symbolised both the coolness and the excess of the Internet, sacked nearly all its employees as its administrators sought a buyer.

Many local IT companies are still reporting widespread losses, slowing sales and a need to restructure their businesses for leaner times. Sausage Software, digital photography specialist IXLA, Keycorp, Reckon, Solution 6 and Techniche are just some of the companies that have revealed substantial losses in recent months.

WHAT HAVE WE LEARNED

Plenty of time has been spent examining the entrails of failed and failing technology companies to try and understand what went wrong. But it's more useful to look at what we've learned from the whole catastrophe and how we can use that knowledge going forward.

As *E-Commerce Times* reported recently, "For all the failures that dot-coms brought to the business world, they also were responsible for some truly astounding technological innovations and drastic changes in accepted business models."

The "vapourware" experience of making promises that couldn't be delivered upon "underscored the importance of adding depth to applications rather than simply offering cliché-ridden value propositions," according to the report.

"'All the dead dot-coms.... had value propositions that sounded like platitudes,' AMR Research analyst Louis Columbus said. 'Every one had a degree of arrogance in them about changing the world, and none of them had real products aimed at real customers.'

But now, according to Columbus, "There is a permanent level of pragmatism out in the market which will never go away. Business is now more focused on results than ever before."

FROM DEATH, NEW GROWTH

As usual, Australian commentators are earthier in their description of the future of dot-coms, comparing them to manure.

BRW columnist Beverley Head writes that "One director who sits on several Australian Internet company boards has begun referring to them as dot-compost rather than dot-coms (because) compost is formed when useless matter decays into a form suitable to be used for fertilising new growth."

She argues that the process of technological composting has been going on for the past 20 years. "The current slowdown in demand follows many others. There was the slowdown in demand for the mainframe, as the market for the mini-computer picked up; there was a slowdown in demand for the mini-computer as client server architectures emerged; there was a shift in demand for workstations as personal computers grew more powerful.

"Each time a supplier decayed, a new and faster-growing company sprang up, fertilised by what had gone before.

"In the past decade, the rapid composting process undergone by the IT sector has been camouflaged by the tide of killer applications, quantum leaps in technology or historical events that forced people to shift to the new generation of equipment faster than they might have liked.

"Today there is little to disguise the slowdown. There are few good reasons being offered for computer users to dispose of what they have installed and replace it with something else. "

NO MORE SCORCHED EARTH

As a result, incremental gains seem to be the order of the day – scorched earth IT implementations are gone.

This applies to large-scale integration projects in merging companies, as well. Wesfarmers, which last year acquired hardware retailing rival

Howard Smith, junked the $21 million SAP software system used by Howard Smith rather than spend the money on upgrading the system and installing it as the platform for the combined business.

Wesfarmers ecommerce director Rich Krasnoff told the *Australian Financial Review*, "I have said over and over, and the same applies with ecommerce, what we are on about is technology solutions to business problems, not technology solutions in search of problems.

"We certainly aren't going to replace a system that meets the business needs just because it's been around a while and because there's hype in the marketplace about all the great things that other systems can do."

That's not to say Wesfarmers is against using new technology. The company is rolling out an electronic procurement system for non-core goods and services across the whole group.

Krasnoff estimated that Wesfarmers spends hundreds of millions of dollars annually in non-core purchases, and he estimated the cost of processing a transaction was reduced from $60 to $10 using e-procurement.

MEASURING THE RESULTS

The necessity to achieve a return on assets means you can't roll the dice the way you were able to two or three years ago when implementing a technology solution. However, the upside is that it imposes discipline on the way you do things.

As Wesfarmers' experience demonstrates, businesses are now using Internet technology only where it makes sense to use it, in areas where it can save them money, as opposed to implementing blanket solutions across every aspect of the business.

Another important point is the fact that Wesfarmers was able to quantify its savings through e-procurement. It is the ability of Internet technology to measure results that will ensure its long-term usefulness.

When it comes to marketing, the emphasis on measurement is a bit ironic, since it was the accurate measurement of results – or, rather, non-results – that made the Internet look expensive to marketers, helping to cause the technology collapse in the first place.

The thing to remember about the Internet is that once you spend money on developing your website, the ongoing cost of developing and keeping leads goes down. In traditional direct marketing and advertising, there is a linear relationship between cost and return. The more you spend, the more you get back. You have to keep spending big to get big returns.

But on the Internet, once you build your site, the cost of a lead goes down. When you've been in the market for a while, your costs drop dramatically. After nearly seven years online, for example, ninemsn's costs of leads are so small they still can't believe it.

Nearly 10 million Australians are active users of the Internet, according to the latest Nielsen-NetRatings figures. The combined television audience is 12 million, which means the Internet audience in Australia is bigger than Channel Nine's audience. Ninemsn has been able to realise the value of its investment in the Internet. By taking a case-by-case approach, so are many other Australian businesses.

ALL I REALLY NEED TO KNOW ABOUT INTERNET MARKETING I LEARNED FROM MY BUILDER

(PUBLISHED IN 2003)

Internet marketing expert Dana Blankenhorn wrote recently on his website a-clue.com, "Innovation doesn't stop during a recession. For the industry leaders, in fact, it accelerates. And that's very good news for all of us.

"The fact is that in a world dominated by Moore's Law (which states that computer chips double in speed and effectiveness every 18 months) all products depreciate as soon as they leave the factory. Recession actually speeds this deterioration of value. You have to innovate in order to survive. You have to actually increase the pace of change, delivering breakthroughs in price-performance, or in the application of technology to solve problems."

It may not look like it at the moment, but many companies are pushing the envelope in developing websites and website marketing. Whereas two or three years ago they tended to be well-funded initiatives, today they are done on the smell of an oily rag – a difference which means they probably have more long-term viability.

Although working on the Internet is more stable and practical these days, I have to admit I sometimes miss the buzz I felt being involved during the Internet boom, that sense of being at the centre of the excitement.

These days, in Sydney at least, the excitement has moved to the property market. The attention once given to web developers is now reserved for builders, the men (and they are all men) who are project managing house extensions and constructing new homes, helping people make a lifestyle upgrade and a sound investment at the same time.

At a party recently, I spent a bit of time chatting to a builder – maybe it was the vain hope that some of his buzz would wear off on me. Anyway, after speaking with him about his work I realised how much his profession and Internet development had in common.

Inspired by Robert Fulghum's book *Everything I really need to know about life I learned in kindergarten*, I've identified eight basic principles that apply to as much to Internet marketing as they do to building:

1. Put in deep, strong foundations. You have to do your groundwork well to give your structure, your website or your marketing campaign something strong to build on. Somewhere down the track you may want to put on a second story, or mesh another website into your original one; preparing now will avoid cracks and heartache later.

2. Find a balance between form and function. A house can have an award-winning design, but be totally impractical. Conversely, one designed solely with practicality in mind can be really ugly and affect resale value. A practical, stark website won't capture the interest of potential customers, but too much style at the expense of substance will make customers say, "It's interesting, but so what?" You need to work hard to blend substance and style so you can capture their interest, then follow up with practical benefits.

3. Allow for future growth. There's no sense tacking a room with a flat roof on the back of your house because it's cheap and easy at the time. A few years later you'll end up tearing it down and putting on three rooms with a full roofline. You'll have wasted the money you spent the first time around and it will be a lot more expensive to build the whole thing than it would have been at the beginning. By the same token, you need to "future-proof" your website so you don't have to junk it and start all over again in a couple of years. Keep an eye on what your needs will be in the future and accommodate them now.

4. Cut corners now, pay the price later. This is also summed up in the agricultural verse from the Book of Galatians: "As ye sow, so shall ye reap." If you "bodgie" something because it's cheap and easy at the time, it will come back to haunt you. You'll look at that paving job you did yourself every day for years and say to yourself, "If only I'd taken the time/spent the money to do that properly…"

The same thing applies to your website. If you spend the money up front on things like usability testing, you'll save later regrets when expensive user-driven changes need to be implemented post-launch.

5. Good communication is essential. If your builder feels he can talk freely with you, ask you questions about things and fill you in regularly on what's going on with the project, you'll get what you want and not what he thought was a good solution according to his logic.

The mobile phone has transformed builder-client communication. The builder I spoke with has a several-hundred-dollar mobile phone bill each month, but he says it's the best investment he's ever made. That's because he can ring his clients on the spot and ask them to make a decision about something rather than him making the decision.

Similarly, if you outsource your web development or web marketing activities such as email or banner advertising campaigns, make sure they talk to you about what you want at every step of the way. That way you'll get what you want and get the results you expect.

6. Getting your hands dirty can both save you money and can help you come up with superior solutions. This relates to the previous axiom. If you take good communication to the next level and get involved in the project – such as sourcing tiles, bath items or lights yourself, doing the painting, building a retaining wall – you can save money. Just as importantly, you'll have more of a sense of ownership about the project. And by asking your builder for advice on where to source items, what colours work best, etc., together you can come up with better ideas than if you relied just on his ideas or just on your own.

The more you get involved with the team who are developing your website or web marketing campaign, the richer the outcome you can achieve. You'll understand the mechanics as well as the objectives better, and it will be integrated into your company's activities.

7. Take the pain in one hit. If you can afford it, put the pool in now, while you've got the back of the house knocked off, instead of doing it in three years' time. That way you can start enjoying the benefits earlier. Sure, there's more chaos and expense today, but there's going to be chaos anyway. Also, your house extension might cut off access to the backyard and make it prohibitively expensive to put the pool in later.

If you commit to an industrial-strength content management system rather than building your own quick-and-dirty one today, you'll make it easier on yourself in the future and start reaping the benefits straight away.

8. Make sure you leave enough money at the end to do the bits everyone notices. Although it's good to take the pain in one hit, you also need to budget carefully to make sure you have money left for the "show" items like paint and landscaping. Particularly if you're looking at resale value, those cosmetic items are important.

On your website, don't spend all your money on a state-of-the art CRM system and then have nothing left for user-centred design so your customers will be happy to use your site.

PORTAL POWER STILL PULLING AUDIENCES

(PUBLISHED IN 2003)

You've heard teachers say: "School would be so much easier without those pesky students." Or salespeople mutter: "My job would go more smoothly if those customers didn't keep interrupting me."

This also applies to technology. Computer security experts are bewailing the fact that the weakest link in IT security is the damned two-legged user. *The Economist* recently reported the results of a British survey carried out by PentaSafe Security that found that two-thirds of commuters at London's Victoria Station were happy to reveal their computer password in return for a ballpoint pen. Another survey found that nearly half of British office workers used their own name, the name of a family member or that of a pet as their password.

By the same token, predicting things on the Internet would be so much easier without dealing with the unpredictable nature of people. If you take a look at the most recent Red Sheriff data on Australian Internet traffic, you'll find that although overall user traffic has increased, the players in the top 10 are nearly all the same as four years ago. Portals (including search engines), Australia's largest corporates (Telstra and the Commonwealth Bank) and Microsoft (no doubt because of all the Internet users downloading patches for MS software) fill all the top spots on the list.

Guy Cranswick from Jupiter research wrote recently in Australia. internet.com's Media Beagle column "the commercial significance of the audience numbers for the major sites (is) that audience numbers are not necessarily a key to success.

"What comes through the entire list is the practicality and functionality of online usage: it's all about finding, learning and doing something."

Getting back to my point about predicting user behaviour, Cranswick also wrote, "The power of the portals is as strong as ever. Only three or four years ago user behaviour, it was believed, would transition into specialty sites (but this) has not occurred at all."

Understandably, the figures show a polarity in time spent online, with news and information portals averaging 10-15 minutes per visit and search engine portals averaging less than five minutes per visit.

Despite the presence of several heavyweight competitors, the ninemsn portal is the only news/information portal in the top 10, and comes in at the top spot, with more than 4 million visits and a reach of more 85% of the Australian online audience each month. Its major direct competitor, the AOL7 web portal, sits in 26th place. In comparison, the AOL/Time-Warner portal is the second most popular website in the US, with a reach of 46% of the US audience, while Microsoft's MSN site, the US equivalent of ninemsn, doesn't even rate in the top 25.

So what is ninemsn doing right here in Australia? As I mentioned, they're not the only game in town. In fact, unlike the traditional Fairfax/Murdoch/

Packer troika in print and broadcast media, ninemsn has literally thousands of competitors on the Internet – and that's just in Australia – it's just as easy for Internet users to visit an international site as it is a local one.

So why is ninemsn so dominant? They offer services such as email news updates and newsletter alerts for the various sites within ninemsn that draw you back to the site. But all good portals do that. Here are a few principles they follow well that have helped build their success, and which contain lessons for emarketers.

First to market – with deep pockets – wins: I know a lot of business experts argue that the second entrant in a market usually wins, because they build on learnings from the first to market and innovating enough to capture interest. To some extent that applies in this case. Ninemsn (or the Microsoft Network, as it was known in those days) came onto the scene after some small online-only players were there, but they had a consolidated effort and threw an enormous amount of money at the Internet to establish their position. A lot of industry observers said they were crazy at the time, but six years down the track, it's hard to argue with success.

Capitalise on what people can't do offline: And the main thing you can't do offline is use email. All portals offer some form of free email, but in the case of many such as AOL7 you need to use them as an ISP to use their email facility. Ninemsn, on the other hand, relies on the world's larges free web-based email. Hands up all those people who have ever had a Hotmail account. Yes, I thought so – plenty of hands in the air. And where do you go every time you log out of Hotmail? The ninemsn home page.

Give people what *they* want, not what *you* want: And when you land on the ninemsn home page after leaving Hotmail, you're not hit in the face by plugs for Channel Nine programs or Packer magazines. The centrepiece of the home page is news headlines. This gives it a newsy, not promotional feel. Users have come to the site because they want to learn about the latest developments in Australia and internationally; chances are they have not come because they want to find out who Don Burke's celebrity gardener is going to be this week (my apologies to both of those people who do, in fact, come the to ninemsn site to find this out). It's all about addressing customer needs rather than reflecting internal structures.

Make it rich, and make it obvious: The ninemsn home page is heavy with lists of menus, each with further drop-down menus hanging off them. Some would argue its too busy, but the effect is that you can see how rich the site is, and you can easily get to nearly every level of the site – no mean feat for a bit site. The content is also varied – as well as articles, there is email, chat, shopping, etc.

Integrate, not segregate - use your offline channels to drive people online, and vice versa: Many shows on Channel Nine prominently promote the fact that you can chat with cast members/experts from the shows after the broadcast, and many of them produce online fact sheets that are promoted on Channel Nine shows and Packer-owned magazines. This is an obvious one, but one that many organizations forget, parking their website in the IT or corporate affairs department rather than marketing, where it belongs.

Notice that I haven't said anything about advertising – the jury is still out on the value of advertising on information or search portals. That's because, like so many other things, counting on people to behave in a rational way is dangerous.

AUSTRALIA FINALLY PUTTING THE BROAD IN BROADBAND

(PUBLISHED IN 2004)

How quickly our expectations change. I've been digging back into past predictions about broadband use in Australia, and found that four years ago ABN AMRO predicted that home broadband connections, then accounting for less than 5% of Internet connections, would soon reach the tipping point and "explode" to reach 625,000 Australian homes by 2005.

In July this year, still a year out from ABN AMRO's prediction, Australia reached 1 million broadband connections. However, analysts say we still haven't reached the tipping point, where uptake takes off and spreads like a virus rapidly through a community. Furthermore, groups like KPMG's technology research group are calling Australians laggards compared to broadband adoption trends overseas.

Mind you, if we haven't reached the tipping point yet, we must be getting pretty close. There were barely 700,000 broadband connections at the end of 2003, which means better than 40% growth in seven months and nearly 100% over the previous 12 months. That's largely due to Telstra dropping its daks earlier this year by introducing introductory plan rates which, starting at $29.95 per month, were lower than the wholesale price they were offering to their competitors.

EMBRACING BUSINESS

Telstra now has an estimated 75% share of the broadband market (when you include other ISPs it supplies lines to), with Optus accounting for 20 percent. They are soon going to be challenged by innovative competitors such as Unwired, which is rolling out wireless broadband in Sydney and has a license that will allow it to eventually cover up to 95% of the Australian population.

The Australian Bureau of Statistics pegs the number of Internet-enabled homes at nearly six million, so there is still a lot of room for growth in home use. Businesses, on the other hand, have fully embraced broadband technology.

A report on the e-business market put out by Telstra's online search arm, Sensis estimated that 43% of small and medium businesses in Australia are accessing the Internet via broadband, while Pacific Internet's Broadband Barometer puts the figure at 52%, up from 23% a year ago.

More than 75% of the businesses surveyed for the Sensis report say broadband has had a positive impact on their business, with reasons ranging from speed of access (69%) and increased Internet efficiency (19%) to frequency of access (18%) and access more applications (12%).

The Broadband Barometer declared that "There remain no killer applications for broadband," with the most widely used technologies among broadband small businesses being firewalls (81%), network security (75%) and LANs (67%). Broadband penetration is significantly higher in metropolitan areas (62%) than rural (24%).

MATURE OR LAGGING?

Ericsson Australia recently released findings from a survey of nearly 2,000 tech-savvy city-dwellers with a strong desire for broadband. The survey revealed that compared to most other countries, Australia is a technologically mature market, with high penetration rates for mobile phones, Internet at home, computer at home, digital cameras and electronic organisers. Ericsson estimates Internet penetration in Australia at 75% in 2004, up from 66% in 2002.

This contrasts another Australian report released at almost the same time by KPMG, which stated that "If broadband is not moved to the high ground of national priorities, we are virtually guaranteed to remain a laggard. The long-term economic consequences of such an outcome could be very severe."

Their justification for such strong statements is quite eloquent, so I'll quote a large slab of it here: "Why does this matter? It is not primarily about cool stuff – even though downloading a CD-quality song in a few seconds for less than a dollar is pretty cool. Rather, it is about the national consequences if, over the next few years, people still waste time and money jumping on planes when they could videoconference but for adequate bandwidth, or die needlessly when telemedicine could have saved them, or still take orders manually in the field only to re-key them back at the office.

"If, compared with our international peers, Australia has lower productivity, a massive imbalance of intellectual property trade with the rest of the world and loss of digital content creators i.e., talent emigration and loss of remote higher education to foreign institutions; then there will be very real, negative consequences for the economy."

On the upside, if we get it right, the benefits are enormous. The Federal Government's Broadband Advisory Group reported last year that "next generation broadband" – the next stage in broadband development, starting to take place overseas, offering much greater bandwidth that current DSL technology – could produce economic benefits of $12-30 billion a year to Australia.

The KPMG researchers wrote that "It is not fanciful to suggest that broadband's importance is comparable to roads or rail when they were developed."

GET CREATIVE

So what does this mean to marketers? Well, the rich media revolution has finally arrived. After what seems like decades of catering to the lowest common denominator of a crappy 56K modem and painfully slow page download times (it has, in fact, only been one decade, but that's 70 years in Internet time), it is finally time to rely on animation, multimedia and video to get your message across.

I predict that we're on the cusp of a creative Renaissance for Internet marketing. Clever designers and producers will now be able to combine the deep information potential of the Internet with the high-impact techniques that have worked well in traditional offline media.

Sadly, I also predict that many lazy marketers will simply put their TV commercials on the Web, without consideration of the distinguishing features and rich possibilities of the new media.

Which category will you fall into?

PERPETUAL BETA: CONTINUOUS IMPROVEMENT, OR NEVER FINISHED?

(PUBLISHED IN 2007)

When you talk about the giants of technology and the Internet, Tim O'Reilly isn't a name that rolls off the tongue. Bill Gates, Steve Jobs, Jeff Bezos from Amazon and Jim Clark from Silicon Graphics and Netscape come to mind before O'Reilly. But he has been a key figure in the industry since before the World Wide Web existed

If you walk past your company's IT department, you will no doubt see some of O'Reilly's work on their desks. His company, O'Reilly Media, is revered by the computer crowd as the publisher of the most practical technical manuals, easily recognized by their use of lithographic drawings of animals on the cover.

O'Reilly has been a strong supporter of the free software and open source movement, which has meant that although he has helped shape technology over the past 20 years, he hasn't made a fortune from his genius like some of his contemporaries.

However, the Irish-American has enjoyed an increased profile in the past couple of years since he coined the term Web 2.0. to describe the transformation of the World Wide Web to a user-driven, continually-updated model of development.

O'Reilly defined Web 2.0 as online applications that offer:

- Services, not packaged software, with cost-effective scalability
- Control over unique, hard-to-recreate data sources that get richer as more people use them
- Trusting users as co-developers
- Harnessing collective intelligence
- Leveraging the long tail through customer self-service
- Software above the level of a single device
- Lightweight user interfaces, development models, and business models.

One of the key concepts to come from O'Reilly's work is "perpetual beta". Wikipedia (which is itself a Web 2.0 application) defines perpetual beta as "a term used to describe a software or system which is always in a testing phase.... Perpetual beta has come to be associated with the development and release of a service in which constant updates are the foundation for the habitability/usability of a service."

What this means is that for services that you can receive over the Internet, such as software and ecommerce transactions, new features are added constantly, as opposed to the old model of huge software releases every few years.

Instead of the Microsoft model, where the company brings out a new version of Windows every three or five years that has a host of new features and has been thoroughly bug-tested, perpetual beta is exemplified by products such as Gmail or Google Maps, where new features are introduced on almost a daily basis.

As O'Reilly describes it: "When devices and programs are connected to the internet, applications are no longer software artifacts, they're ongoing services. How long have we been waiting for Longhorn? Whereas Amazon, eBay, Google, they just roll in new features, unsure whether they even want them... therefore don't package up new features into monolithic releases: rather, fold them in on a regular basis, such as Flickr, Google's often very rich new features, deli.icio.us. So if you're not already thinking this way: operate as if you're in perpetual beta."

Jim Morris, on his Software Product Management blog, writes, "When software delivers its service over the web, we can do business differently. The ability to fix the software without distributing updates is helpful. Since interactions between users and servers go over the net, they can be recorded and replayed. All of these things can be exploited to support much better bug analysis and performance monitoring. Software just got easier!

"An application should evolve to better serve its purpose. So, aside from providing a service that attracts users, it should be gathering information about what else the users might want or need. Thus, the tough decisions that product managers make about which features to include in the next release can be made in a more informed way.... features can be tried on small subsets of users until they prove dependable and popular."

PRACTICAL APPLICATION

That's fine for technology giants such as Google, Amazon and eBay, but how does this apply to ordinary businesses?

It means changing the way you think about your customers and how you do business. If there is any service you can provide to your customers online – even in the form of information – you can get it out there straight away. You can also gather feedback easily and immediately, in order to make changes to your products. This could offer a huge competitive advantage.

However, you have to resist the temptation to just throw something out to your customers because you haven't put enough thought into it and you want your users to tell you how to finish it.

In his Signal vs. Noise blog, Web designer Jason Fried writes that the problem with forever keeping products in beta is that it fosters fear among developers that their work will never be good enough to be "final."

"I think it's kind of ridiculous," said Fried. "I think that people are maybe ashamed of their products and are worried about releasing something that's not perfect. It feels like it's almost an excuse. They're putting something out there and saying, 'Use this, but if it's not perfect, it's not our fault.'"

Lane Halley, a member of several social-networking services, told Wired that such free services have the prerogative to do whatever they want. But she also argues that companies must be careful with what they make available to the public.

"The thresholds of calling something a beta have dropped through the floor," said Halley. "As soon as a company has something they're ready to show, they'll put (it) out there.... I think that's a very dangerous tendency because you end up with products that are not clearly defined."

Francesco Mapelli writes in his www.mapelli.info blog: "What scares me is that the use of the term 'beta' seems to have misled some companies... giving the impression that they can build buggy services, put a colorful beta logo and think they are cool and 'so Web 2.0'".

"The architecture of a service can be in perpetual beta, because users can provide important feedbacks and feature requests, and companies should add and remove tools and services depending on the feedback and their interests. What can (and should) be in perpetual beta is the way small pieces link together, not the small pieces itself.

"Web 2.0 is like LEGO. You have small, simple and colorful pieces, and you build platforms and services with this pieces. You can build something that changes every day, but the small pieces you use must be solid."

ENTERPRISE ANYTHING

(PUBLISHED IN 2011)

I'm quite sure Google rewrote the book on best practice. I remember our technical director arguing with an enterprise architect about how eBay had scaled their platform. The architect had one opinion, eBay had another. The issue was simple – in a very short space of time eBay had rewritten best practice.

Enterprise guys have grown up with IT best practice and that means projects that don't finish, projects that run years over time and over budget. These people think that if you buy programming at $60 per hour offshore, that's better than paying $150 per hour to do it here. They don't take the time and the number of people into account.

A few years ago Senator Alston, while he was Federal Communications Minister during the Howard years, had a website that ran $20 million over budget – that's right, not a $20 million website, but $20 million over budget. The web industry said the job should cost about $500,000, to allow for the $400,000 overhead of dealing with government. The government had no experience in this kind of thing so they gave it to an enterprise that had no experience and ran way over time and budget. Enterprise, I'll bet, was mentioned on every PowerPoint slide that sold that job.

Real web people know the rules have been rewritten. New languages and frameworks make it quicker and easier to do things. Enterprise software is great for the time and materials consultants who are happy when it takes three weeks to set up a platform. Web guys, however, don't want to take three weeks because we want to get this job over with and move onto the next one because it might be more interesting and we might learn something new.

At the moment, that something new doesn't mean traditional enterprise solutions – it means going open source, where the source code is freely available to any developer who wants to fit it to their needs.

Open source means you can test something, measure it and prove it without spending a fortune. The current Gerry Harvey "GST-dodging online retailers are killing my business" furore will not be solved by enterprise software; it'll be solved when the current big daddies of retail invest in getting started and spend the next five years focused on building and optimising these channels.

ENTERPRISE MEASUREMENT

Google Analytics means you can measure your activity. Enterprise analytics vendors are quick to sell the shortcomings of Google. Meanwhile, many of their products can take a year to implement correctly and require a complex team to run and manage. Most non-enterprise analytics tools give you insights from day one and after 12 months you're streets ahead of the enterprise that has done nothing but implementation.

ENTERPRISE TIMELINES

Enterprise CRM implementations that never quite finished. We used to spend six weeks setting up Vignette across a suite of servers. Now, simpler open source software like Alfresco enables this to be done in hours, leaving the bulk of the budget to actually getting the product right.

When an enterprise platform does not work as advertised, you have two choices: Spend a fortune fixing it or start again using something else. There are no refunds in this area.

Enterpise CRM is another matter. Many of these systems and can now be replaced with SASS products like Salesforce.com or open source products like Sugar . Again, you can put time you put into actually doing are so much more valuable than the time put into implementation.

ENTERPRISE PROCESS

Enterprise process that requires every stakeholder's involvement and sign-off has delivered very little. Projects are not better off, they take longer costs more and there is no incremental improvement in the outcome.

I recently left my name and telephone number of an enterprise marketing automation website. They haven't gotten back to me yet, but in the meantime a small SASS provider that I approached got back to me immediately. We've already closed the deal and it's for a large corporate client. If you snooze, you lose, enterprise software.

Time is the most important part of any IT strategy. If you're slow, the landscape has changed before you get your project finished. Speed to market is all that counts. Say no to enterprise anything and get started with something that works.

OPEN SOURCE: IT NOT ONLY TASTES GOOD...

(PUBLISHED IN 2011)

I interviewed Alfresco Software's regional sales manager, Barry Costin, on the topic of open source vs. enterprise software for a recent podcast.

Why tackle such a seemingly technical topic for a marketing blog (albeit a marketing and technology blog)? Why should we care about the intricacies of software? Well first, let's start with some background for marketing-softened brains. Here's a definition of open source software from How Stuff Works:

"Most software that you buy or download only comes in the compiled ready-to-run version.... It is extremely difficult to modify the compiled version of most applications and nearly impossible to see exactly how the developer created different parts of the program. Most commercial software manufacturers see this as an advantage that keeps other companies from copying their code and using it in a competing product. It also gives them control over the quality and features found in a particular product.

"Open source software is at the opposite end of the spectrum. The source code is included with the compiled version and modification or customization is actually encouraged. The software developers who support the open source concept believe that by allowing anyone who's interested to modify the source code, the application will be more useful and error-free over the long term."

The open source, free software movement started back in the 1980s with products like Linux, but things have moved on from the early days of unshaven Dungeons & Dragons-playing geeks playing around with software code. Today there are many commercially-available products that call themselves open source. As Barry Costin explains, "Open source means the software itself is free – you can take code in free mode and beat it into shape. Companies don't charge for a licence, but you charge for maintenance and support, indemnity and warranty – all the things related to ongoing relationships."

Anyone who has ever been involved in an enterprise software project, particularly on the procurement side, knows that upfront licence fees can be up to $500,000, with implementation, maintenance, etc. on top of that.

"Customers are now pushing to get better value for money," Barry points out, which means they're turning more and more to open source solution providers, which make their money out of maintenance and support rather than licence fees.

"It's cheaper, but there are also other advantages," Barry says. "You're the master of your destiny - you can change it, there's a more rapid rate of innovation, you can tap into a large community of developers working to improve the product."

The advantages of enterprise solutions are that there's an upgrade path, someone to blame when it goes wrong, and that you're not tied to a particular developer.

In the past, open source was synonymous with untested, buggy software. But modern-day solutions offer the same level of support and indemnity as enterprise solutions, but are lighter and easier to work with, and without the high upfront fees. In fact, competition from open source companies is forcing enterprise solution providers to lower their fees and adopt a pricing model similar to open source companies.

"The game is up for enterprise software companies," Barry says. "Venture capital investors are now looking almost exclusively to software-as-a-service (SAS) or open source companies. Customers know how to orchestrate very good deals with software vendors."

Years ago, you used to bend your company to fit around a commercial software product. Now you bend the software to suit your company. This allows you to do custom development, agile development, while controlling risk. You can develop in bite-sized chunks rather have than Battlestar Galactica projects."

He adds that, "You can get so much more mileage out of your IT dollar than you could even just five years ago – it's just insane. Moore's

Law [which says the cost of technology keeps halving] is working on the hardware side – software is changing at equally rapid rate."

Bite-sized, iterative development means you can focus software solutions on what you need today, test and adjust. As Barry says, ". If you're just buying a car to go the shops and buy some milk, don't buy a Rolls Royce. Good enough these days is good enough; it doesn't need to be gold-plated."

So, to get back to the question at the beginning of this piece – why should you care about a topic like open source vs. enterprise software? Because you can save lots of money, time and heartache, and you can quickly build a solution that can be easily improved upon.

A better, more flexible, faster solution that costs less? Yes, please! As the blues band Canned Heat used to say: "It not only tastes good, it's good for you, baby! So get it – and don't forget to boogie."

USER EXPERIENCE/ USER DESIGN

SCHIZOPHRENIA ON THE WEB

(PUBLISHED IN 2001)

There's a war of words – or, more precisely, words and pictures – raging across the Internet. And depending on which side wins, ecommerce could develop in radically different ways during the next decade.

As the Internet has evolved from a network of academics swapping research papers into a business tool, entertainment medium and purveyor of community, the pony-tailed crowd has begun to take a keen interest in its development. This has caused no end of annoyance to the practical-minded usability experts who emerged during the early grey-background-with-blue-underlined-links stage of the Web as the rulers of the online world.

In their eyes, the Internet is all about content and making it easy for people to find what they're looking for. And these days, it's not as easy as it used to be. As Jakob Nielsen, poster boy for the usability crowd, says, "There are 20 million sites on the web, and nearly all of them are awful."

When people design web sites, he complains, they rely too much on their own perceptions of style and aesthetics, without giving enough thought to the average user, dialling up on a slow connection and lacking the sophisticated aesthetic sensibilities of the design professional.

As a result, users get frustrated and give up trying to fight their way through the heavy graphics to find what they really want, and become more and more ruthless in their judgement of sites.

Nielsen and his colleagues have developed 222 general principles of design for ecommerce sites, ranging from keeping logos to a minimum to always putting a search box on the home page. While no website has been judged perfect by Nielsen's standards, Amazon.com and Google.com came up as two of the best-designed sites on the web.

On the other side is a new crop of Web designers who are loading sites up with animation, colour and abstract concepts and arguing the experience is the brand and the brand is the experience. Their poster boys are people like Gene Na, co-founder of Web design firm Kioken, who says, "What the client sometimes doesn't understand is the less they talk to us, the better it is. We know what's best."

Na and his adherents are making headway on the Web. More and more sites are relying on weighty (in terms of download time) devices such as Flash as opening animations or to drive their navigation. According to Jupiter Media Metrix, Flash is the tool of choice for 54% of advertisers, with the next best Enliven and Real Audio at 27% each. Macromedia, meanwhile, claims that 92% of Web users can view Flash content without downloading a player, countering Nielsen's arguments against the use of plug-ins.

GOING TO THE DOGS

Each side thinks it's the only one with the answer. But a school of experts is emerging who argue that the answer to good web design lies somewhere in between. This school is represented by people like Vincent Flanders, author of *Web Pages That Suck* (based on the website of the same name), a man who can't decide whether the Web is like dogs, sex or California.

He writes, "What Jakob and, to a lesser extent, the art community forget is that you can't judge a website by such logical standards as usability or creativity. You have to judge a web site like a dog show.

"At a dog show you have the different breeds broken down into several groups. Inside each group you judge the dog by the standards for that breed. You don't judge a poodle the same way you judge a bulldog. It's the same with the Web."

He also writes, "The web is like a strip mall in Monterey, California. Monterey is a beautiful seaside town with loads of educational facilities, beautiful architecture, great restaurants – and a few, ugly strip malls. That's the real Web – moments of beauty surrounded by ugliness and crass commercialism.

Finally, revealing his sympathies with the usability camp, Flanders writes, "Don't confuse Web design with sex. Web designers are confusing the Web world with the real world. In the real world, foreplay is mandatory. But in the world of the Web there's no place for foreplay. It gets in the way. The Web is 'Wham, bam, thank you ma'am.'"

Despite what usability experts and designers each say about their position, website design is not black and white. And the sooner website builders work that out, the sooner the Web can reach the next level in usefulness.

INCREASED RETURNS – BY DESIGN

(PUBLISHED IN 2002)

Nothing is so simple that it cannot be misunderstood.
– **Freeman Teague Jr.**

> If people can't figure (website usability) out, they deserve to go bankrupt. – **Jakob Nielsen**

The US usability research firm User Interface Engineering has reported that half of all transaction on the web fail, because users fail to reach their goal. For ecommerce transactions, the figure is 65%, while only 30% of users find what they need on a site using search.

According to Forrester Research, consumers these days are finding little value in business website content. Web users visit manufacturers' sites to research products – and 84% of these active prospects expect those sites to provide the best product information. But only 45% offer relevant and complete content. Forrester reports that manufacturers fare worse than most, "typically offering glitzy images in place of useful information."

The Forrester report goes on to say, "Even though consumers, business customers and site executives underscore the need for a great user experience, most Web efforts don't deliver it. Nearly 150 site reviews conducted by Forrester since December 1999 show that most sites fail to support user goals. User experience matters, but most sites don't deliver."

WEBSITES ARE PRODUCTS, TOO

Most businesses view their website – correctly – as a marketing tool, another method of promoting and selling their products. I've been advocating this view for as long as I've been in website development, emphasising the importance of integrating a website with a company's marketing armoury and overall business processes.

But while all that is true, it's also important to view your website in another way: as one of your company's products.

And as with any product, attention needs to be paid to how your website is designed and how easy your customers find it to use. But with websites, that's not happening, particularly in Australia.

One of the biggest reasons this happens is that technology is driving business use of websites, instead of the other way around. Scott Berkun, design and usability for Microsoft, compares this to the chicken and the egg dilemma, asking, "Which came first, the user or the technology?"

He writes that, "A good craftsperson in any trade understands that people will consume their work, and every decision is made with that type of person in mind. Software or Web development is no different. Makers of automobiles, CD players and appliances all have the same challenge of balancing engineering, business and usability, except they've been doing it a lot longer than we have."

ACHIEVING ROI FROM DESIGN

(PUBLISHED IN 2002)

There's no point making changes to a website unless those changes are going to provide more business benefit than what they cost to implement. Fortunately, there is an increasing amount of data on what types of changes are likely to generate the most return. Here's the consensus of opinion on how to increase return on investment from website redesigns.

Listen to your users, not your designers: Web consultant Dean Allen, writing for the online graphic design website A List Apart, described how he taught a class for third-year students at a website design school, and found that by the time he encountered the design students, "their education had plainly focused away from what I consider the primary goal of communication design: to make vital, engaging work intended, above all, to be read – to use design to communicate."

He included in his article what he calls "an entirely incomplete list of things a non-illiterate designer should know before being a designer". The list includes:

- That the practical value of white space towers over its value as a design element.
- That the deep symbolism of a design decision, referring perhaps to a treasured memory of the designer, is irrelevant to the person attempting to glean something from the work.
- That the physiobiology of reading is one that demands easy points of exit and entry.
- That overstating the obvious can be effective, but not all the time.

Put a human face on technology: Jef Raskin, one of the creators of the Apple Macintosh and author of the recently published book, *The Humane Interface*, argues in favour of immersive interfaces, where you aren't really aware of the computer or the physical operations you need to do in order to accomplish a task.

He writes, "Interfaces must be designed to accommodate our ability to pay conscious attention to only one object or situation at a time. When we perform multiple tasks simultaneously, unless all but one of them are being performed habitually, they will interfere with each other and we are more likely to make errors."

Make it easy to search and navigate: Jakob Nielsen, the Web's highest profile usability expert, says that if your site search box is not on your home page, move it there. Jef Raskin agrees, stressing the importance of having an effective search methodology covering both the site and the web as a whole. He also advocates developing navigation "as applicable to finding your way around within a picture or memo as within a collection of images, documents, or networks; a method which makes use of inborn and learned human navigational skills."

Be prepared to spend money up front to get it right: Nielsen says that to get the most out of a website, spend about 10% of a project's budget studying how people will use it. Spending the money upfront, he says, will yield far greater revenues later because the site will be easier to use.

Despite the benefits, Nielsen said barely 20% of companies he surveyed test Internet projects before putting them online. Nielsen said a key reason companies don't do user testing is that the department required to pay for it – information technology – is generally not the one that reaps the rewards.

Be prepared to spend money at the end of the process to measure whether you got it right: Forrester Research argues that determining ROI from website design changes is difficult because most companies don't even attempt to measure the effectiveness of their changes. Many companies are spending between $100,000 and $1 million on redesigns, and while most track high-level business results, like revenue or customer satisfaction, few have any sense of which specific design changes paid off – if any.

This is often because, as Forrester says, "The real experts get put in a box." Usability specialists like Nielsen can demonstrate ways of improving users' ability to find information and complete orders. "But even gurus like these can't offer many examples of the *business* impact their changes made. Why? These expensive consultants typically get packed off once their design work ends – long before the new site is up and running. With no time or incentive to measure, most don't."

If a design change isn't aimed at increasing profit, don't do it: Forrester says projects that focus on adding brand imagery from TV and print ads are usually graphics-intensive and will frustrate all users except those on broadband, so they should be cut.

If you haven't already, move to a content management system: The bigger websites get, the more it makes sense to use page templates. They not only provide a consistent user experience, they make it easy and cheap to make both global and individual changes to a website. Although they can be costly to implement, they can provide measurable payback quickly.

Taking the product design approach to websites focuses site redesign work on solving specific business problems and puts the emphasis on making a site easier for customers to use. With specific goals in mind, it's also possible to measure the success of website changes and ensure a return on website investment.

IT'S TIME FOR IRRATIONAL EXUBERANCE

(PUBLISHED IN 2003)

The list of things people would rather do than swallow barium and submit to a full body x-ray is, frankly, quite long. But would you believe that one of the things people would prefer is a colonoscopy, a 90-minute procedure that achieves the same result as the barium treatment but involves the even more distasteful and uncomfortable process of shoving a camera up, um, well, a part of the body most people like to keep hidden?

More importantly, what on earth does this have to do with website marketing?

Daniel Kahneman, a psychology professor at Princeton University in New Jersey, won the 2002 Nobel Prize for Economics for his work on consumer behaviour, which has blown the idea of the "rational consumer" out of the water.

The rational consumer, who weighs costs and benefits before spending his or her money on a consumer good or service, has been a cornerstone of economic theory and marketing for most of the last century. Kahneman, on the other hand, says that consumer decision-making is influenced more by the desire to be happy and comfortable than by rational thought.

Guy Cranswick reports in *E-Media Marketer* that Kahneman's most famous experiment involved 700 patients undergoing a colonoscopy. For half of the patients, he lengthened the procedure by leaving the colonoscope inside the patient for an extra minute. The key was that during this extra minute, it wasn't turned on, which, while uncomfortable, is more pleasant than having it buzzing away inside you.

Cranswick writes, "Mysteriously, those who had the extra minute had a consistently more favourable reaction afterwards and were much more likely to elect for a colonoscopy again, rather than the less intrusive barium meal and x-ray. Kahneman deduced that the last level of experience establishes what is retained of the whole experience."

Kahneman's takeaway from this experiment was that people's strongest memory of an experience is what happened at the end, and if they had a happy experience (or at least, a less unhappy one), they would be more likely to repeat it. I can't explain how colonoscopies lead to buying behaviour, but Kahneman did so sufficiently to snag the Nobel – the first time an economics prize was awarded to a psychologist.

There are a few lessons here that can be applied to website development and marketing. First and most obviously, if your customers find that dealing with your business through your website is as pleasant as a barium enema, they're likely to tell you to turn your website into a colonoscope and use it to examine the inside of your company's digestive system. The easier you make it for customers to do business with you, the happier they will be and the more business they'll do with you.

The second lesson – much harder to quantify – is that there are many ways of gauging your return on investment (ROI) on your website that are intangible and not necessarily rational, and they are as important as the things you can easily measure.

Log file analysis or direct online sales on their own don't paint the true picture of your website's contribution to your company profits. A Forrester report on car manufacturers' websites, for example, found that although customers weren't buying cars direct from the carmakers' websites, those sites were proving to be important tools in the car buying process.

At end of the car selection process, car buyers were more likely to visit manufacturers' websites – and then make the deal. Nearly 30% of people visiting a carmaker's website bought a car within six weeks – the same rate as people visiting the showroom. A corporate website today is a "hot lead".

People are relying on the Internet more and more for reliable information before making a purchase or other decision. According to a recent Pew Internet Life survey, consumers use the Web as a central, primary information source. Whether hunting for product information or info about government, health care or news, 97% of Internet users expect to find the information they seek online.

As Sean Carton writes in *Clickz*, "The Web has become so well known as an information source that 64% of *nonusers* report they'd expect to find information they need online. If information about your product or service isn't available online, it might as well not exist at all."

"Exactly how and when the Web gets worked into the buying decision process may be ultimately unknowable... and ultimately unimportant. The point is, consumers are going online expecting information they want will be there.

The Pew researchers concluded that, "For business, it is clear that an online presence is important, regardless of whether a business actually sells its wares over the Internet. If a store provides product information online, but doesn't sell products at its website, nearly half of all those surveyed said this would still make them more likely to go to the physical store to buy the product."

The practice of applying broad, qualitative measurements in conjunction with easily measurable quantitative ones is already well established in offline marketing, with many industries conducting "brand health" surveys that compare actual sales with measures such as propensity to buy – that is, a customer's attitude towards buying something.

There can often be a wide difference between propensity to buy and actual sales. Actual sales tend to be driven more by retail advertising and price, while propensity to buy is driven by factors such as brand advertising, history with a product and trust in a company and its products. These sorts of factors can be heavily influenced by a company's online presence.

So can the cost-saving side of a website. US researchers the Yankee Group estimate 66% of companies still determine website success by measuring traffic, followed by new customer acquisition at 34% and revenue generation at 23 per cent.

A Yankee analyst point out that cost-reduction measurements that scored lower in the survey, such as fewer customer service calls (20%) and customer service cost savings (19%), are what companies should be looking at. The analyst told *Darwin* magazine, "It's way past the time to just look at 'eyeballs' or 'stickiness' – look deeper and find out how the website helps the sales department or marketing department cut costs, learn more about customers to provide better customer service and so on."

By incorporating a range of "irrational" ROI measures along with the standard traffic and ecommerce measures, businesses can get a broader idea of the "real" ROI of its website. Setting clear, customer-focused business objectives will make the process much less painful than a barium enema – for businesses and their customers.

THE RIGHT PERSONA FOR THE JOB

(PUBLISHED IN 2003)

How well do you really know your customers? Do you know the most important ones by name, where they live, what their attitudes are toward your company and your products or services? If you knew some of them that well, would it make a difference to the way you design your product/service, your marketing campaign, your website?

I think most marketers would agree that the answer to that last question is yes. Most marketing depends on intimate knowledge of customer types and customer groups, but not of individual customers themselves. There are excellent reasons why you don't get down to that level of knowledge, the main one being that any single customer is bound to have some idiosyncrasies that would skew the results of marketing campaign design. No customer can be typical enough to design an entire campaign – or website – around.

But a new type of framework for Internet design has arisen that turns that sort of thinking inside out. Appropriately enough in an era of cloning and computer simulations, it involves "creating" a typical customer from scratch. The concept, called personas, is revolutionising website design overseas, and is likely to soon have a strong impact in Australia.

Simply put, personas are detailed descriptions of user archetypes that represent market segments – a composite description of a real person who represents a primary customer segment. These personas are then used to drive website designs and changes.

EDUCATING REBA

Personas are built following interviews conducted with representative customers from each of a business's key target segments. A profile of the persona is created that treats this archetype as a full-blooded person, with a name, gender, family history, personality, attitude, even a photograph – although the person is totally fictional.

The concept was developed by software designer Wayne Cooper, who started out play-acting how a particular user would interact with a software program. He wrote a best-selling business book *The Inmates Are Running the Asylum* in 1998 in which he describes how personas work for interactive design. The book "was intended to alert managers to the problems inherent in designing software for use by non-engineers," according to Cooper.

It doesn't sound like rocket science – that's because it isn't, although as Wayne Cooper says, "Personas, like all powerful tools, can be grasped in an instant but can take months or years to master."

Forrester Research in the US has championed the use of personas in Website design. Forrester analyst John Dalton says, "Personas must provoke a sense of empathy for the design team. Personas must therefore have the attributes of a real person, such as name, face, age, job, home address, family, and ambitions." Dalton adds that a key feature is a 'day in the life' vignette of the persona, including their interaction with the company's web presence.

It's important to limit the number of personas you use to two or three types representing only the key customer segments – and to create a persona who represents the toughest customer in that segment.

Forrester analyst Harley Manning writes that "Sites created for everyone are doomed to satisfy no one. By designing primarily for the most-demanding personas, design teams can develop focused scenarios that please the most-demanding users without disappointing the rest."

Forrester's Bruce Temkin says personas limit the focus of design, but "that's exactly what they're supposed to do. Design efforts that try to satisfy all customers produce interfaces so dense with controls, content, and function that even expert users struggle to manage them. Design personas... empower design teams to focus their energies on essential content and function."

"Design personas," Temkin says, "help companies make informed, fast design decisions. By creating a shared, vivid picture of target customers' behaviours, project teams can better evaluate how to satisfy customer needs.

"The impact is less scope creep from unwanted and unnecessary features, faster consensus across the team, and none of the pitfalls from self-referential design. Rather than suffering through endless debates about design priorities, teams can settle disputes with a pointed question: 'What would Reba want?'"

Kim Goodwin, design director at Cooper Design, says a key difference between a persona and a customer type is that "Personas represent behaviour patterns, not job descriptions."

Bruce Temkin says, "Designing a marketing campaign is different than designing a Web site or an IVR interface: people just absorb a message, but they *interact* with a system. Marketing-centric customer profiles suggest the promise to make to customers. Personas reveal how to build an interactive system that delivers on that promise."

PAINTING SCENARIOS

Forrester has dubbed the process of applying personas to particular situations to test a website "scenario design". It asserts that scenario design can have a demonstrated financial impact on business websites. As Harley Manning says, "Ecommerce integrators with the expertise to create personas and designs based on scenarios will have an easier time showing ROI for their designs."

Manning points out that knowing your customer segments well is a prerequisite for building personas and conducting scenario design. "Many firms still lack the critical user information necessary for successful scenario design."

John Dalton agrees, saying, "The primary is the persona that poses the most stringent constraints on the design – he's the toughest customer to please. But designer beware: If the team can't identify a primary, then the project is in trouble. Failure to locate a primary means that either the application is trying to do too many things at once or the research has failed to collect sufficient data regarding customer attitudes and goals."

Although Manning argues that personas are "critical to the process of scenario design", he writes that even among its biggest customers Forrester "seldom encounters companies that use personas."

He asks, "Why do companies shy away from investing in these essential tools? They fear that they'll lose customers if they don't anticipate and satisfy *every* potential goal for each of their users, they fret that they'll pick people who aren't typical, and the pressure to show return on investment is high."

But, he concludes, "None of these objections holds water. Site managers could make faster, better decisions through a deeper understanding of users' online behaviour, technical expertise, and detailed goals – the kind of understanding that personas provide.

"Valid personas reduce work while saving both time and money over the course of a project because they serve the most important market segments, they simplify design decisions, and they help avoid costly rework.

PREDICTIONS – GENERAL

SHOPPING IN THE NEXT MILLENNIUM

(PUBLISHED IN 1999)

When making predictions about where shopping and ecommerce are headed, there's a good chance you'll get it totally wrong. After all, 10 years ago the World Wide Web didn't even exist, and now ordinary shoppers are spending billions of dollars a year buying things via their computers, while traditional retail behemoths are quaking in their boots over the threat of competitors that have no stores.

Will something equally transforming and unanticipated arise during the next 10 years? Maybe. But despite that possibility, as we reach not only the end of a decade but a century and millennium to boot, the temptation to look forward is too great. So, here are my predictions about how ecommerce will reshape the shopping experience over the next several years.

Plastic will win over virtual cash. This is an easy one; it's on the cards (pun not intended) already. According to BizRate.com, only 3.6% of online buyers have used a digital wallet and 58% are not even familiar with the concept (and this is a US survey, so the figures are likely to be even lower for Australia).

It's a matter of logic. Why invent a complicated new payment system when an existing one can do the job? Online security concerns are being addressed daily. There is arguably more credit card fraud perpetrated via credit card numbers provided over the phone and via fax than online. Most people already have credit cards, so it makes sense that they use their credit cards online as they do for other purchases.

The ultimate in one-to-one marketing -- one-to-one selling -- will bring about the biggest change to the way people spend their money. When personal computers gave ordinary people the ability to design their own publications, it was predicted that everyone would become a publisher. Now, the rise of online auctions means that everyone can become a merchant.

As Meg Whitman, CEO of Ebay says, "We are enabling a kind of commerce that didn't exist to any extent before, and that's person-to-person commerce." Auction sites have spontaneously generated the types of communities that corporates have tried and been largely unable to create.

That's why big players such as Microsoft and, in Australia, Fairfax and News Limited, have been quick to join the bandwagon started by new players such as Ebay and Stuff. Like shopping malls before them, online auctions will put the experience in shopping experience.

People will spend more money online, but they won't necessarily spend more money. This one follows to a certain extent from the previous prediction. People aren't spending more money because of the availability of online shopping; they're just spending it differently.

Only 6% of ecommerce revenue in 1999 will be in the form of incremental sales; that is, sales that wouldn't have otherwise happened, according to Jupiter Communications. Even by 2002, when

Jupiter predicts consumer ecommerce sales will top US$41 billion, incremental sales will only amount to US$3 billion, or just more than 7% of revenue. That means that nearly 95% of online sales have been cannibalised from traditional avenues - retail stores, direct mail, telesales, etc.

Consumer spending is not a voluminous vat into which online commerce revenue is poured to add to the total spend. It's a pie, one that is only growing as fast as overall consumer spending grows. As the size of the piece given to ecommerce grows, the size of the other pieces will shrink. (See next prediction).

The successful online 'etailer' will be the one who blends offline infrastructure with online expertise. The World Wide Web didn't exist 10 years ago, and now we're already hearing about 'post-Web retailing'. US online research firm Forrester Research recently released a report describing post-Web retailers, those sellers in the next millennium that will understand their customers well enough to be able to anticipate their needs.

They will centralise all available information about shoppers, gleaned from online and bricks and mortar activity, and will be able to identify their most profitable customer segments. They will then identify and adopt new product categories that will appeal to those profitable customers. True post-Web retailers will move beyond personalisation, the current Holy Grail of online marketers, to anticipatory selling.

The store of the future will have both an online and physical component, and is likely to look like a cross between a department store and a category killer specialist outlet. Forrester consumer ecommerce analyst Seema Williams, for example, predicts that "The post-Web winners will be brand-name merchants with a strong offline infrastructure and the ability to establish an online retail gateway."

Distribution, not online technology, will be the biggest stumbling block to the success of ecommerce. It may be easier to turn on your computer and search the web for the best bargain than it is to let your fingers do the walking or drive to the mall. But if it takes too long or is too expensive to receive your online order, people won't use it. Etailers like Amazon and grocery site Peapod have already realised this and are building warehouses to make sure they can properly stock orders. This is the trump card of traditional retailers - they already have a distribution chain in place.

Traditional retailers are going to look more like online retailers - and online retailers are going to look more like traditional retailers. This relates to the previous prediction. Just as traditional sellers will not survive into the future without embracing online sales in some form, online-only businesses will not survive unless they have an offline infrastructure in place for distribution and marketing. Look for more mergers and partnerships between online and offline companies.

Window shopping will continue to be a popular online ecommerce activity. Research is indicating that while approximately 33% of Internet users have bought items online, twice as many claim they have shopped on the Web but bought elsewhere. That gap will start to close up, but it will probably never be closed completely. The Internet provides the best way to do comparison shopping, and people who will remain set against buying online will still use it to help make their choices.

Traditional stores will still exist - and still be popular at Christmas. Last year the Christmas season was a boom time for online shopping, and this year's predictions are for online holiday activity to grow even bigger. But no matter how convenient it is to shop online, some people are too disorganised to buy all their stuff online ahead of time (see earlier point about distribution). 24 December will always be one of the busiest shopping days of the year.

TOO MUCH MEDIA, NOT ENOUGH TIME

(PUBLISHED IN 2000)

Predictions of the shift in media use to new and emerging technologies are frighteningly large. What will the impact be on the media, advertising and marketing professions? Here's an informal SWOT analysis on what is likely to happen over the next 10 years.

STRENGTHS

The customer: In the early days of the Internet content (information) was king. As the online media become more focused on ecommerce, the truism for the next decade is that the customer is king. We're moving into an era where people feel empowered and are expecting to be in control. This presents a huge challenge for advertisers and marketers (see weaknesses, below).

All companies will tell you they're customer-focused, but there's always an element of manipulating customer response. With online shopping, you have customers empowered with new tools to help themselves – what will they end up buying when you can't manipulate them? They'll buy the best product at the best price. This forces businesses to change their approach from punting things at people to saying, "What can I do for you?"

Simple solutions: Amidst all this whiz-bang technology, the best solutions will be the simplest. And online, it doesn't get simpler than email. Already bigger than the postal service in most countries, email will grow from 10 billion messages this year to 35 billion by 2005 (International Data Corporation figures).

WEAKNESSES

WAP: I use my mobile to communicate with friends. The day I choose to let Proctor & Gamble contact me – forget it. People will simply not put up with getting a message on your phone in Grace Bros. telling you about a sale in your shirt size on level four. People don't want to be sold, they want to buy.

Closed technology: Proprietary systems and technology that dictates to the user will not survive. Telephone companies, even Microsoft and Apple now understand this.

Ad agencies: Three years ago, if anyone had an opportunity to kick a goal online, it was ad agencies. They had customer relationships, they had the imprimatur to develop those relationships in a new medium, and they dropped the ball. Of the big Web developers worldwide, only a few have their origins in an agency base. Agencies have been good at TV, but the rise of the Web has meant the decline of TV and the decline of agencies.

Why haven't they made the transition? I think it's the fact that agency work is fundamentally centred around the cult of the individual – the power of one creative idea. To do the Web, you need a team involved.

Shopping malls: Online, people operate more with shopping lists and less by impulse. Although malls won't die off, customers generally won't be as influenced by nice-looking displays.

Marketing in general: New media technologies are making marketing a lot harder. I don't watch TV, I listen to the radio in the car during the 10 minutes it takes me to drive to work, I read mainly US magazines, I get most of my news and information off the Web, and when I get DM, I grab the bill and throw the rest in the bin. There are more single people around now. They don't say "Chicken Tonight, got to have it." They are getting more cynical – they know that driving a certain car doesn't give you a bigger penis.

OPPORTUNITIES

Consulting firms: It's proving difficult for ad agencies to get staff, because they're going to consulting firms who have grabbed the online world with both hands. McKinsey will have the power that used to be accorded to ad agencies.

B2B: If companies are saving 18-20% off the cost of items by using business to business portals, that means they can either buy more things or buy different things than before – as long as, as stated before, they are the best products at the best price.

Broadband and interactive TV: Although these two are developing independently, their offerings are too similar for both to survive. I predict interactive TV and broadband will merge their production mechanisms, though they will still probably fight over the mode of delivery.

ECRM/personalisation: Customer relationship management and one-to-one marketing will grow and thrive, but current predictions of its take-up are overblown. Users will want to feel in control.

THREATS

Stable market populations and shrinking markets: By the end of this decade online technologies should achieve market saturation in Western countries. While people with Internet and broadband access tend to spend more time overall consuming various forms of media, there are only 24 hours in a day. You can only watch so much standard free-to-air TV, pay TV, HDTV, Internet, broadband, WAP, interactive TV. There will be a lot of market fragmentation.

At the same time, production costs for these new media are much higher than equivalent costs for normal TV and radio (It's said around the industry that aging Australian TV stars will need to spend three times as long in make-up hiding their wrinkles when digital TV comes online).

How can you justify higher costs for a smaller market? Digital cultural imperialism from large markets such as the US and even China pose a real threat to the Australian media over the next 10 years.

Metabrowsers: Another example of the power of the Internet being tilted in favour of the customer, metabrowsers are, as Andy Goldberg calls them in the *Sydney Morning Herald*, "browsers on steroids" that enable users to copy and paste content from other websites into one personally-designed web page. Obviously, if people can choose what they want to go on a page, they're going to leave out the ads. As online analyst Jakob Nielsen says, "It's just a temporary phenomenon that the Web is ad-driven."

That begs the question, how will it be driven? It will be driven by the user, the customer. Only those businesses that truly become customer-focused will survive beyond the end of the decade.

DEEPER, YET SHALLOWER - THE PARADOX OF DIGITAL ENTERTAINMENT

(PUBLISHED IN 2006)

Predicting the future can be a risky exercise. It's all too easy to end up with egg on your face when reality unfolds over time. I remember as a child the Jetsons-like predictions that by the beginning of the 21st century we'd all be flying around in personal flying cars and wearing shiny one-piece outfits. Instead, we're still traveling in our cars (albeit in a lot more comfort) underground in expensive road tunnels rather than in the air, while our kids wear fashion better suited to a 19th century gothic novel than a science fiction tale.

When it comes to speculating about how we will entertain ourselves in 20 years, I think it's useful to look at where we are along a continuum;

by looking at how things have changed in the past 20 years we might get some idea of what to expect in the next twenty.

Twenty years ago, the world was still assessing the overwhelming effect of television on all forms of entertainment. Attendance and revenue for live events from cinema to theatre to sporting fixtures were all in freefall because it was more convenient and cheaper to watch it all on TV. MTV was starting to affect radio playlists and record sales, as video became an essential part of a song's appeal.

The World Wide Web, meanwhile, was just a gleam in scientist Tim Berners-Lee's eye two decades ago. No one was predicting the digital revolution that would sweep through the entertainment industry and which is now overtaking the seemingly unbeatable impact of television itself.

When you think about it, although the Jetsons scenario has not come to pass, we've gone through a hell of a revolution in entertainment in the past 20 years. The influence of the World Wide Web and its developments over the past decade, from ordering tickets online to sharing songs through Napster to building social networks and sharing videos on MySpace and YouTube, is changing almost every aspect of entertainment and is in the process of usurping the seemingly impregnable position of television. And it's not over yet: digital progress will mark the biggest changes to come in entertainment, arts and sports in the next 20 years.

One of my favourite phrases is, "The revolution will not be televised". Although it was not coined with the Internet in mind (it was the title of a song by black activist and singer Gil Scott-Heron back in 1971 about the American Black Power movement), it has become a mantra for me in my work in the online media. It doesn't mean that digital is going to kill television, but it does mean that television is no longer in the drivers' seat. Narrowcasting, not broadcasting, will be the dominant paradigm in entertainment over the next 20 years.

Digital doesn't just mean the Internet. Digital technology will become the norm within the traditional entertainment media as well as the new media. Movie theatres are already moving to digital – all major movie chains are expected to convert to digital film by the end of the decade. Films will be available on computer files or small cassettes rather than huge canisters of film, and every film shown in cinemas will be an "answer print" – only one generation removed from the original (mass release films are currently second-generation prints). There will be no problems with wear and tear and no "spatial jitter" as the film moves through the projector.

Digital movies are expected to save the studios more than a billion dollars a year in distribution costs. They will also make it considerably cheaper to show the same movie on multiple screens at the same cineplex. This will further reinforce the current blockbuster mentality – think 25 sessions per day for Pirates of the Caribbean at your local Greater Union, but little else.

That's not to say that more blockbusters are a bad thing. The incredible special effects in movies such as Superman Returns (such as the buckling metal in the passenger jet as Superman saves it from crashing) are arguably worth the money being spent to achieve them, as long as there is quality storytelling and not just special effects. Blockbusters will continue to become more high-tech and more realistic over the next two decades, making it difficult for low-budget films to compete.

While logic would dictate that this would spell the death of arthouse cinema, I think it will give it a new lease on life. Arthouse cinema will become the "long tail" of movies (more about the long tail a bit further on). As the Internet increases the profile of niche products, arthouse cinema will thrive as it caters to those niche players who still want a group experience. If you like arcane French cinema and you have to look for it now, in the future you won't have to look so far. Movies pushed out of cineplexes by bigger blockbusters will move to the independents. Cinema will become both deeper and shallower at the same time.

Back at the cineplex, digital technology will be used to extend "live" experiences beyond the stadium and opera house. In the US, a company called National Cinemedia is piping rock concerts from major arenas into cinemas in towns that major bands would never visit. You can't smoke dope or dive in the mosh pit, but at $15 it's a hell of a lot cheaper than a concert ticket. At the other end of the culture scale, the New York Metropolitan Opera has begun transmitting live performances to movie theatres and broadcasting more than 100 live performances over the Internet or on digital radio in an attempt to expand its audience.

The blockbuster mentality has affected theatre as much as it has the movie business in the past 20 years – though Andrew Lloyd Webber is as much to blame as television. This trend is likely to continue, as the cost of putting on a show becomes even higher relative to other forms of entertainment. People will only spend $100 on a ticket if they know they are going to see a proven winner like Les Miserables or The Lion King. Families have limited budgets, so it will be harder and harder to justify going to live theatre.

Although local amateur theatre will continue, non-blockbuster shows by metropolitan theatre companies will have to change the way they work. They will turn to the Internet to keep their production costs low and to expose themselves to a wider audience. YouTube will become the new Broadway, hundreds of times over. Of course, the easier it is for theatre companies to build a profile online, the easier it is for bad theatre companies to build a profile. The theatre will become both deeper and shallower.

If you thought Napster and iTunes represent the pinnacle of personalized music, just wait until you see what will happen online over the next 20 years. By aggregating music choices of millions of users, new online music services will take you beyond what you know you like and move into predicting what you will like. Pandora Radio, one of the first applications in this area, allows you to set up your own radio station, where you tell the program what type of music you enjoy and it selects a menu of songs that include not only the artists you have selected, but music of a similar genre and sound. Another site, last.fm, scans your computer or iPod and makes song choices for you based not only on what you have downloaded but, more importantly, which ones you actually listen to. It then uses an algorithm to create a playlist that mixes your collection with other songs it predicts you will enjoy. And all of this is free as long as you are prepared to listen to some ads amongst the music. As *Wired* says, "Spyware never sounded so good".

The stranglehold that record companies and radio stations have had over new music was shaken first by Napster and Kazaa and then by iTunes. Artists like indierocker Beck are taking the next steps in the disintermediation between artists and their fans, providing a model of the disparate nature that music releases will take in the future. His most recent album, *Guero*, was released in several forms – first as a work in progress leaked onto the Internet, then as a studio CD, followed by a DVD with seven extra songs and professional music videos, and an album of remixes. He even produced homemade music videos filmed by his family members and loaded them onto YouTube. This multimedia approach to music will become more pronounced in the next 20 years.

Unlike television or even other Web sites, YouTube makes it easy for bands and fans to establish a dialogue through video. It's easy to email video links to friends, or paste links onto individuals' websites, blogs, or MySpace pages. Posting tribute videos and comments is also easy. The lead singer of OK Go, a band that has built its eclectic reputation through clever videos on YouTube, told *Wired*, "It sort of provides an infinite information jungle gym for our fans, one that's always growing and morphing," constructed as much by the fans as by the band itself.

Another way bands will use the Internet to connect with fans in the future is through virtual concerts. Singer-songwriter Suzanne Vega – or rather, her computer-generated avatar - recently performed a virtual concert on the SecondLife website, where users create alter egos and interact with each other in an alternate society. Visitors to the site could hear her live performance and watch a rather clumsy digitally created representation of the event; Vega's avatar struggled to hang onto her virtual guitar and her mouth was rarely in synch with the music. One reporter who attended the concert said, "all this is pretty goofy. But it's likely the birth pains of a new age of music performance."

And although the rise of iTunes is threatening to kill album sales in favour of singles, services such as TuneBooks will ensure the album concept won't die. Customers download TuneBooks along with an album online and then navigate a QuickTime-powered trip through pictures, videos, credits, and lyrics. It will soon be available in a mobile phone version, and will no doubt lead to more eclectic ringtones, ones that will hopefully sell more than the Crazy Frog (speaking of the shallow end of the music pool).

One feature of the digital revolution that is having a strong commercial effect across the entertainment industry is the concept of "the long tail". Coined by Chris Anderson, the editor of Wired magazine, the long tail refers to the fact that back catalogues for just about anything are able to be sold cheaply via digital means, so old and obscure titles are enjoying a renaissance. While bookstores make 95% of their income from a maximum of 100,000 titles, Amazon.com's 3.7 million titles are all equally available, and half of its income comes from books outside the top 100,000.

Anderson cites the example of *Into Thin Air*, a recent book about a mountain-climbing accident that, through Amazon's "also recommended" engine, breathed new life into a 15-year-old title on the same topic,

Touching the Void. Sales of the older book ended up overtaking the newer one, and it was made into an award-winning docudrama that drove further sales for the original book. This story is being retold over and over across videos, DVDs, music and all forms of archived entertainment.

What about TV – how will it address the digital challenge of the next 20 years? The current teething pains around digital broadcasting will be settled over time and interactivity will become a much more important part of the TV viewing experience. That means more reality TV, a concept that is driven more by the opportunity to participate in decision-making than the promise that Australian Idol will find the next Elvis or Big Brother will uncover the next game show host.

The relative cost of producing quality drama and sports programming for TV will continue to go up as audiences go down. A couple of things will happen as a result. Name actors and sportspeople may not be able to command the same salaries as they do at present. Many new shows will be available only via services such as YouTube, filmed using low-cost equipment to suit the less polished nature of online video.

Of course, the problem with everything ending up on YouTube is that our tolerance for low production values will grow. We're already getting used to pixellated photos and mobile phone photos – even on the news. We will become pickier and less picky at the same time as we make the time to watch some quality drama on TV, but also consume (usually in bite-sized chunks) the fast food options available online.

As bandwidth becomes less of an issue, video will be added to more online services, making them more like (badly done) television. In the SecondLife vein, a research group at Northwestern University has developed a text-to-speech engine called Buzz that identifies most-searched topics on popular blogs and uses a virtual host to read the information out loud. The next step will no doubt be David Tench summarizing and commenting on the web's top blogs.

Gaming is also going down this route, with games that allow you to put your own face on your on-screen counterpart. Games will also use applications like VirtuSphere, which was designed for military training and lets users roam around simulated environments without being limited by physical barriers. A wireless headset senses the movements of virtual explorers, letting them navigate lifelike situations inside the freely rotating orb. Its makers claim you can run, jump, or crawl your way through anything that happens on-screen without hurting yourself on any actual obstacles.

One guarantee for 20 years' time is that we will have a much wider range of choice in entertainment. The great unknown is what will be worth choosing.

NEXT: THE NEED TO LOOK FORWARD IN DIGITAL MARKETING

(PUBLISHED IN 2011)

No sooner had LinkedIn nudged $US9 billion on the day of its IPO this month, then the talk started that this was the start of Tech Bubble 2.0.

Although the valuation for LinkedIn was high – nearly 40 times earnings – warnings of another bubble are premature. The digital industry has developed dramatically since the heady days of the first tech bubble. Back then, fuel was supplied by dot-coms burning through cash faster than venture capitalists could supply it while generating no revenue, their untested business ideas nevertheless floating for hundreds of millions of dollars.

LinkedIn, on the other hand – the first major social media property to float on the stock market – has 100 million members, has been around for eight years, earned nearly $250 million last year and is on track to earn $400 million this year. Hell, they're even making a (small) profit!

What else is different now? John Frankel, a partner at FF Venture Capital, told The Huffington Post that in 2000 and 2001, "It was all about eyeballs and audiences – and not generating real revenues." He argued that "technology has evolved by leaps and bounds since then... everything from location-based services, cloud computing, mobile, gaming, couponing, local advertising and small business adoption."

Ten years on from the bursting of that bubble, the Web has developed to the point where no one would argue that it is anything but a paradigm-changing feature of business. The world has changed, and in the next five years it's going to change even more; businesses need to prepare for that change.

So, what do you need to consider when looking to leverage the power of digital? These days you need to move beyond the mechanics, such as selecting the right content management system, learning how to track ROI against your other marketing activities, or deciding whether or not to launch a Facebook page – all that is fait accompli.

You need to look forward so you can understand and capitalize on the trends that are going to become tomorrow's digital opportunities.

Despite what local digital practitioners say, Australia is not at the forefront of digital development. We are at least 18 months to two years behind what's happening in the US and Europe, and that's where we need to look to plot our future.

Who should you listen to? There are a number of established experts who you can follow on Twitter or read their blogs – people such as former Apple executives Guy Kawasaki and Steve Wozniak, marketer Seth Godin, search veterans John Batelle and Bruce Clay, or social media commentator Chris Brogan.

All of these experts have at least one thing in common; they've all appeared on a podcast show that's on the must-listen list for digerati around the globe.

The quirkily titled DishyMix podcasts are the brainchild of Susan Bratton, an ebullient and hyper-connected veteran of the Silicon Valley digital scene who has interviewed 200 digital marketing experts for her show over the past five years.

When you speak with Susan and listen to her recent podcasts, a few themes that are worth noting by Australian marketers emerge: driving traffic, driving engagement, and the art and science of persuasion. All of them are underpinned by the technology of measurement and the changing face of consumer behaviour.

DRIVING TRAFFIC

When you read about driving web traffic, you automatically think of search, right? Not necessarily; search is important, but you also need to move beyond search and understand how to drive traffic from paid sources, using tools such as demand-side platforms and performance media.

A demand side platform is a software-driven system that allows digital advertisers to simultaneously manage multiple ad exchange and data exchange accounts. Demand-side platforms can offer efficient ad exchange media buys by offering an easy-to-view, cost-effective, automated approach to ad exchange buying.

Why should marketers care? Because demand-side platforms are starting to automate nearly every part of the media-buying process. That means traditional media buyers are going to have to skill up or give way to a new breed of technology-driven buyers. The way media has been bought – and the people who drive it – are going to change. Demand-side platforms offer a viable option for dealing with the frustrating fragmentation that's now inherent to the online media space.

Performance media is another example of online marketing automation that is gaining prominence. Essentially, it's a generation or two beyond pay-per-click – you pay for an online campaign only once you've achieved your desired outcome, which can be a click, an enquiry, a lead or a sale.

Performance media marketing campaigns are usually delivered via the major online publishers either via banner display ads, rich media or email direct marketing.

DRIVING ENGAGEMENT

Of course, digital media extends way beyond paid messages. It is focused on distributing experiences. As Henry Jenkins, the author of "Convergence Culture", says, "The key is to create something that pulls people together and gives them something to do." That's something that social media sites like Facebook do very well.

The challenge for marketers is to leverage the opportunity offered by tools like Facebook and Twitter to engage customers and deepen their relationship with your brand. A lot of digital marketers in Australia will tell you that they know all about social media marketing, but when you look at what's happening overseas, it's clear that we have a lot to learn.

Susan Bratton says that social media marketing "is more than just posting your latest commercial on YouTube. It is a fundamental move in marketing strategy from interruption and push to invitation and engagement. It is turning a prospect on to a brand idea enhanced by the surrounding context."

So how do you drive engagement with customers? To start with, the term itself is probably a misnomer, because it implies that brands are in control of the journey. As the seminal online marketing tome "The Cluetrain Manifesto" pointed out more than 10 years ago, "markets are conversations". We now get what that means: consumers are in control of marketing conversations.

Susan Bratton describes this turnaround in marketing activity and power as "going from scheduling media and blasting out to everyone, to consumers, through consumer-generated content, being equal to what we can do as marketers." Instead of companies with customer databases, customers now have databases of the companies they want to deal with. You may want to drive engagement, but your customers are the ones behind the wheel.

PERSUASION

The Social Network, the Hollywood film about the early days of Facebook, provides a great snapshot from the infancy of social networking, when it was more important to build the community than to commercialise it. But times have now moved on; while Mark Zuckerberg & Justin Timberlake – uh, I mean Sean Parker – steered away from advertising in those early days, in 2010 Facebook generated nearly $2 billion in advertising revenue, a figure that is expected to top $4 billion in 2011.

That reflects the maturation of social media marketing in the past few years. Both customers and brands are now using social media in more meaningful ways than they did even five years ago. Companies don't just own a Facebook page, they are actively using it to persuade people to buy their product – and social media users not only accept that, they welcome it as part of their conversations online.

Today, new businesses are being built around partnering with customers online to develop insights that deliver value to both customers and businesses. Bazaarvoice, for example, builds customer communities for companies that seek out customer reviews and other conversations to drive their business decisions.

In a world driven by measurement and 24-hour cycles of communication, this type of activity provides a better understanding of what it takes to persuade people to do business with you - not through traditional methods of shouting and pointing out features and benefits, but by listening and offering things of relevance to customers.

Looking at what's happening in all of these areas overseas should help Australian businesses to get a handle on the best ways to make use of the Web. It also provides comfort that, unlike the first tech stock bubble, this time around companies have learned how to make money out of the Web.

ABOUT THE AUTHORS

Simon van Wyk is an Australian digital pioneer, serving as the managing director of HotHouse Interactive for more than 20 years after it was founded in 1994. At HotHouse Simon developed some of Australia's first corporate websites and performed web consulting, strategy and development for Australia's top corporations and government entities, including Toyota, Telstra, Optus, Coca Cola, McDonald's, Vodafone, Philips, E*Trade, Australian Business Online, Austrade, NSW Office of Information Technology, Parenting NSW, HCF, Vertical Markets, Aventis Pharma and Deloittes. A regular contributor to publications including B&T, *Marketing*, marketing & e-business, internet.com Australia and *The Sydney Morning Herald*, he is a sought-after speaker at conferences in Australia and overseas, refreshing in his frank statements about both online and traditional business. Simon is currently the managing director of the digital strategy consultancy Blue Road Group.

Dr. Ray Welling has worked as a journalist, editor, managing editor, publisher, content marketer and creator, strategy consultant, writer and lecturer. One of Australia's most experienced digital content specialists, he was HotHouse's first content director in 1995. Ray has a BS in journalism from Northwestern University's Medill School of Journalism in Chicago, an MA in mass communication from Macquarie University and a PhD in ebusiness marketing and management from the University of Sydney. Winner of the inaugural Australian Innovation award at the Pharmaceutical Innovation, Research, Marketing and Education Awards, he consults on digital content strategy and teaches digital marketing, media/marketing convergence, public relations and social media at The University of Sydney Business School and at Macquarie University, where he recently managed to get an article about Kim Kardashian published in an academic journal!

www.ingramcontent.com/pod-product-compliance
Lightning Source LLC
Chambersburg PA
CBHW020831210326
41598CB00019B/1865